Unemployment and
Technical Innovation

Unemployment and Technical Innovation

A Study of Long Waves and Economic Development

Christopher Freeman
John Clark
Luc Soete

Greenwood Press
Westport, Connecticut

Library of Congress Cataloging in Publication Data

Freeman, Christopher.
  Unemployment and technical innovation.

  (Contributions in economics and economic history ;
no. 50)
  Includes index.
  1. Technological innovations. 2. Unemployment.
3. Economic development. I. Clark, John, 1946-
II. Soete, Luc. III. Title. IV. Series.
HC79.T4F74   331.13'7042   82-1113
ISBN 0-313-23601-1 (lib. bdg.)   AACR2

Published in the United States and Canada by
Greenwood Press, a division of Congressional
Information Service, Inc., Westport, Connecticut

English language edition, except the United States and Canada,
published by Frances Pinter (Publishers) Limited

First published 1982

Copyright © 1982 by SPRU, University of Sussex

331.137042
F855

Library of Congress Catalog Card Number: 82-1113
ISBN: 0-313-23601-1
Printed in the United States of America

CONTENTS

ACKNOWLEDGEMENTS

We acknowledge a particular debt of gratitude to the Social Science Research Council for a research grant which financed the work on which this book is based. This programme is continuing and we hope to develop both the analysis and the policy side of the work during further research. We also wish to thank many colleagues in the Science Policy Research Unit (SPRU) for their advice, assistance and criticism and particularly Jackie Fuller, Keith Pavitt, Austen Albu, Giovanni Dosi, John Thógersen, Peter Senker and Paul Gardiner. We are grateful to Joe Townsend, for help with the indentification of major innovations and to Harshad Pitroda and Sally Wyatt for computational help. We are grateful, too, to colleagues outside the Unit who contributed ideas and made helpful comments, especially Dick Nelson, Rod Coombs, George Ray, Roy Turner, Christopher Saunders, Stanley Metcalfe and Charles Cooper. We are grateful to Klett-Cotta, the publishers of *Konjunktur, Krise, Gesellschaft: Wirtschaftliche Wechsellagen und soziale Entwicklung im 19. and 20. Jahrhundert: Proceedings of the Bochum Conference on Long Waves in Economic Development* (see Petzina and Van Roon 1981), to the Kiel Institute of World Economics, to *Futures* (special issues in 1981), to Philip Allan, publishers of '*Some Economic Implications of Microelectronics*' (see Freeman 1981), to Macmillan Press Ltd, publishers of *Controlling Industrial Economies* (ed. Frowen 1982) for permission to draw upon earlier papers. All of these papers have, however, been extensively revised and reworked for the purpose of this publication.

Finally, we would like to thank Jenny Bentley and Linda Gardiner for their unfailing and patient secretarial help in enabling us to complete our work.

INTRODUCTION

The debate on the effects of the micro-processor (the 'chip') has shown an extremely wide range of views about technical change, employment and unemployment. Some commentators tend to look upon this and other technical innovations as a major source of unemployment, whilst others stress their beneficial effects in generating millions of new jobs. Probably all would acknowledge that technical change both destroys old jobs and creates new ones, but the emphasis on one or the other side of the process varies widely. One school of thought tends to assume that there is an inherent equilibrating tendency, which keeps the two tendencies in balance. Whilst not denying the existence of such equilibrating tendencies, our analysis in this book also points to sources of disequilibrium and to the links between technical innovation and fluctuations in the economic system.

The high levels of unemployment recently experienced have led to renewed interest in these problems of long-term structural change and technical innovation — problems largely neglected by both Keynesian and monetary economics. In commenting upon this neglect, Jewkes and his colleagues (1958) observed: 'Future historians of economic thought will doubtless find it remarkable that so little systematic attention was given in the first half of this century to the causes and consequences of industrial innovation' (p. 19). They explained the neglect in terms of the inherent difficulties in measuring and understanding technical innovation, the specialisation of science and technology, and the preoccupation of most economists with other issues which appeared to them more urgent and more susceptible to analysis — such as unemployment and the overall management of the economy.

We would certainly not underestimate the first of these difficulties, and we are only too well aware of the inadequacy of our own attempts to analyse and measure the flow of innovations and their economic consequences. However, the fact that a problem is difficult to understand and to analyse should not deter us from making the attempt if it is important enough. And it is our contention that the problem is of great importance. Moreover, as will become evident, we do not regard the problems of unemployment and overall management of the economy as separate from the problems of technical innovation, so that we do not see them as alternative pre-occupations, but rather as part of the same pre-occupation. Economic analysis and policy making that lacks a technological dimension is simply not good enough to cope with our present problems.

This is not to say that the economics profession has been underestimating the importance of technical change. Such a criticism would be unfounded and totally unjustifiable. One of the things on which economists of all persuasions are in agreement is the great importance of technical change for the long-term growth of productivity and for raising standards of living. However, because of the great complexity and diversity of technical change and because of its unfamiliarity, there has been an understandable tendency to treat it as a 'black box' and make simplifying assumptions about it. After a while these simplifying assumptions tend to be taken for granted; for example as part of the traditional *ceteris paribus* clause in much economic analysis. It is, of course, essential to make some simplifying assumptions, as otherwise no progress could be made through the impenetrable jungle of detailed observations. However, it is also necessary to remember that any model or abstract representation of the behaviour of a system can only be as good as the assumptions that are fed into it, and that if these are seriously in error, misleading results may emerge.

For some short-term analysis it may not matter very much that some doubtful simplifying assumptions have been made, as their effects may be relatively insignificant over a short period. However, if the analysis is extended to the long-term then serious misjudgements may result. It has now become obvious that some assumptions about economic behaviour which appeared to be valid in the 1950s and 1960s, and gave apparently good results in forecasting models at that time, were no longer valid in the changed circumstances of the 1970s and still less the 1980s.

It is the contention of this book that the simplifying assumptions about technical change in much economic analysis can obscure the real processes of change rather than clarify them, if we are considering long periods such as half a century. For example, it is often convenient to assume a constant rate of technical change or that technical change is 'neutral' in some sense, i.e., as having neither a 'labour-saving' nor 'capital-saving' bias. A common assumption made in many neo-classical growth models (see, for example, Solow 1969) was that although capital stock per person would tend to increase steadily as a result of technical change, it would do so in such a way as to preserve a constant ratio of capital to output thus maintaining conditions for a steady long-term full employment equilibrium growth path. In the long period of high growth and relatively full employment after the war such an assumption appeared plausible, but (as we attempt to show in detail in Chapter 8) it does not provide a satisfactory description of what has been happening since. Such questions are far from being hair-splitting academic controversies; they have an important bearing on the prospects for employment in the 1980s and the scale and type of public and private investment needed to improve these prospects.

Just as a realistic technological dimension is fundamental to a proper understanding of the behaviour of the economy, we also believe that an historical and institutional dimension is vital. The development of industrialized economies cannot be reduced to statistics of the growth of GNP, of industrial production, of capital stock, investment, employment etc., valuable though these statistics undoubtedly are. Underlying these statistical aggregates are the growth of entirely new industries and technologies and the decline of old ones and many social and institutional changes in the structure of industry and government. An account of the growth of nineteenth century economies that ignored the railways, the joint stock company or the steam engine would be seriously inadequate and the same would be true of an account of the twentieth century that ignored the electronic computer, the oil industry and the multi-national corporation. This means that statistical analysis must be complemented by economic, social and technological history, if it is to illuminate the real processes of change which we are trying to interpret. This is why we have rejected a purely econometric approach to the problem. On the other hand a purely descriptive anecdotal historical analysis is inadequate without some attempt to measure the overall trends in the economy and the principal components.

Our method, therefore, is one of 'reasoned history'; we attempt to interpret the statistical evidence of changes in the international economy over a long period in the light of a discussion of the incidence of major technical and organisational innovations and their assimilation in the economies of the industrialized countries. After a brief introductory chapter in which we present the basic statistical facts about long-term trends in unemployment and discuss some problems of measurement, definition and theory of unemployment, we take up the issue of the occurrence of radical innovations and their impact on the economy. This approach to economic development is associated, above all, with the name of Joseph Schumpeter and it will be evident from Chapter 2 that we owe a great deal to him. In this chapter we discuss the relationship between innovations, long-term structural change and fluctuations in investment and employment. Schumpeter insisted that opportunities for profitable technical innovation were very unevenly spread over time, being subject to explosive bursts as entrepreneurs realised the possibility of exceptional profits and growth arising from new combinations of technical and organisational change.

Following Schumpeter we stress some of these discontinuities in the process of technical change. We argue that major inventions and innovations have been unevenly distributed in their incidence, both over time and in terms of their effects on the various sectors of industry and services. They lead to the rapid rise of entirely new industries and technologies and to the decline and disappearance of old ones. Such processes have an international dimension, as well as a regional dimension within countries, and they may be far from smooth and painless in their impact, since they are essentially disequilibrating. There is no warrant for complacent assumptions about self-correcting adjustment mechanisms, which would lead rapidly and automatically to 'compensating' new employment for all the jobs lost through technical change. On the contrary, even in terms of non-Schumpeterian economic paradigms, there is little to justify such complacency, and more reason to expect some of the adjustments to be long-term and far from automatic.

We then discuss in Chapter 3 the 'bunching' or 'clustering' of innovations and the possible relationship between such clusters and long cycles of growth and stagnation. A pioneer in this field has been Gerhard Mensch, who in his book on the 'technological stalemate' (1975) propounded the view that depressions accelerated the introduction of radical innovations, which consequently tended to cluster in deep depression decades − such as the 1880s and the 1930s − and provided the impetus for subsequent phases of strong growth and prosperity. Although we criticize some aspects of this theory rather strongly in Chapter 3, we would like to acknowledge our debt to Mensch and his colleague, Alfred Kleinknecht, for their pioneering and thought-provoking work in this field of innovation studies. Our differences with them relate not to the importance of radical innovations and clusters of innovations, but to the social and economic factors which influence the timing of such clusters and their reciprocal influence on the behaviour of the economy.

In Chapter 4 we further develop our own approach to the clustering of innovations and their relationship to the growth of new industries and the economy more generally. We stress the importance of the 'diffusion' process and the way in which a series of further innovations are generated as a swarm of imitators move in to invest in a new technology, attracted by the exceptionally high profits achieved or anticipated by one or more of the pioneers. Unlike Mensch we stress also the role of advances in basic science, whether in physics, chemistry or biology, and of social, managerial and organisational changes in triggering and facilitating clusters of basic inventions and innovations. The bunching of technically related 'families' of innovations may not only give rise to several new industries but may simultaneously or subsequently affect many other (existing) industries — as for example in the familiar contemporary case of micro-electronics. We therefore discuss the rise of 'new technology systems', rather than just new 'industries', in order to capture the full scope and complexity of these developments.

In Chapters 5 and 6, in line with the basic approach that has been outlined, we attempt to illustrate the argument with two examples of new technology systems that have been particularly important in the post-war growth cycle — synthetic materials and electronics. These examples show that one or several major new technologies are capable of imparting a sustained impetus to the economic system lasting several decades. However, as economies of scale are exploited and the technology is to some extent 'standardized', a more capital-intensive phase of growth sets in. Following the entry of a swarm of imitators, profits are gradually 'competed away' and then the opposite process of concentration sets in, accompanied by a 'shake-out' with the elimination of some competitors. During the early period of rapid growth and imitation there is a tendency for new firms, including small ones, to move in and for employment to expand rather rapidly; but as a technology matures, competitive cost pressures, standardization and labour-saving technical change associated with economies of scale tend to predominate. The pattern of investment changes from capacity expansion to rationalization and gives rise to a lower rate of employment growth, or even to a reduction of employment.

We are, of course, only too well aware that the two examples which we discuss certainly do not encompass the totality of economic and technical change which has taken place since World War II. But, for the purposes of this book, we have confined ourselves to these two illustrative chapters. However, in Chapters 7 and 8 we extend the analysis to discuss how far the more generally available statistics of industrial production, productivity, investment and employment, particularly for the UK and USA, support the type of analysis which we have advanced.

Chapter 7 discusses the variations in growth of output, productivity, prices and employment between the principal branches of manufacturing, contrasting the fast and slow-growing sectors. It attempts to relate those variations both to the direct effects of technical change and to the changing pattern of consumer demand. Chapter 8 discusses trends in capital investment and the growth of capital stock in the principal industrial countries. The tendency for smaller increments of employment to be associated with each new 'vintage' of machinery and plant may mean that massive new investment would be required to achieve a return to full employment, or in other words that unemployment could persist as a serious problem because of 'capital shortage'. The implications of this analysis are that the problem of unemployment must be approached not only in the context of short-term manipulation of demand or the money supply, but also in the long-term context of technical and structural change and the associated international trade competition. The pattern of new investment that is feasible is highly constrained by this intense international technological competition.

In Chapter 9 we take up the question of the changing *locus* of international technological leadership with all the associated consequences for patterns of trade, investment and employment. It is obvious enough that countries such as the UK, which were the leaders in the early phases of technical innovation and industrial growth, are no longer in that position today and are suffering from some of the pangs of adjustment. The leadership that passed to the USA and Germany in the third and fourth Kondratiev cycles (i.e., the waves of expansion associated with the internal combustion engine, electric power, the heavy chemical industry, and other clusters of radical innovations) is now challenged by Japan. Such changes depend not only on scientific or technical leadership but also on the capacity to initiate those social and organizational changes that facilitate the widespread adoption of the new families of innovations. This leads on to the brief concluding discussion in Chapter 10 of the policy implications of our analysis.

From this outline of the contents of the book it is evident that it has two major themes: unemployment and the process of long-term technical change. The first issue — unemployment — is discussed relatively briefly and mainly at the beginning and the end of the book in Chapter 1 and Chapter 10, whilst Chapters 2 to 9 are largely concerned with the dynamics of long-term economic growth. This balance, however, does not reflect the balance of our concern, but simply the fact that it is necessary to understand some aspects of long-term structural and technical change in order to cope with unemployment. We believe the problem of unemployment to be the most serious confronting the industrialized countries in the 1980s and analogies with the 1930s are by no means far-fetched.

It must always be remembered that Hitler's accession to power in 1933 was directly associated with the effects of the Great Depression. Support for his party remained relatively small, even immediately after the run-away inflation of 1923, but his vote soared from 1930 to 1932, when mass unemployment provided a fertile breeding ground for his doctrines and thousands of recruits for the storm troops and other paramilitary formations. The riots in English cities in 1981 were a further salutary reminder of the potentially explosive psychological and political consequences of large-scale unemployment associated with inner-city decay.

In a project closely related to our own Jahoda (1982) has concluded that the destructive psychological consequences of unemployment for the individual are more severe than the alienating effects of the less pleasant forms of employment. She has also concluded that the harmful effects of unemployment in the 1980s are potentially as severe as those of the 1930s, in spite of the intervening changes in social welfare arrangements. She was one of the first to study the destructive social psychological consequences of prolonged unemployment in Marienthal, a village in Austria, in the 1930s and her views must command great respect.

If the analysis in this book has any validity it will not be easy to avert some of the adverse consequences of large-scale unemployment in the 1980s. However, it is our hope that by focussing attention on the long-term problems associated with technical change and economic development, we may contribute in a small way to the amelioration of these difficulties.

Unemployment and
Technical Innovation

1 THE DIMENSIONS OF THE UNEMPLOYMENT PROBLEM AND THEORIES OF UNEMPLOYMENT

The purpose of this introductory chapter is to provide some basic statistics on unemployment and to outline some theories of the relationship between technical change and employment. The chapter is divided into three parts: first, data are presented on the level and structure of unemployment; secondly, the notion of 'full employment' is discussed to provide a context for assessing the magnitude of the problem implied by the data; and thirdly a brief discussion of causes of unemployment, with particular reference to technical change, is presented. The overall intention is to provide a background for the more specific theoretical ideas and empirical evidence presented in later chapters.

1.1 Levels and structure of unemployment

There is little need here to dwell at length on the fact that measured unemployment has shown a persistently high and generally upward trend in most OECD countries over the last decade. Table 1.1 summarises the situation prevailing in a number of countries in selected years in the post-war period and in the Great Depression of the 1930s, in both relative terms as a percentage of civilian labour force (Table 1.1(a)) and absolute levels (Table 1.1(b)). Comparison with the unemployment levels of the 1930s shows that the situation is not yet as serious as it was fifty years ago in most countries, especially in the leading industrial countries — USA, Japan and Germany. However the deterioration in the overall position since the 1960s is clear as is the rapid deterioration in 1980–1. There are no grounds for complacency, particularly as many economists believe that serious difficulties will persist well into the 1980s.

Some commentators maintain that data such as these, taken at face value, do not accurately reflect the problem facing developed countries. Such arguments fall into two main groups. First, it is sometimes suggested that the data suffer from serious, even overwhelming problems of measurement. Large numbers of people are included who should not be — those who, for example, are incapable of work, or are 'voluntarily' unemployed. Secondly, from a more theoretical perspective, some schools of thought would regard

Table 1.1(a)   Unemployment levels, as a percentage*

	1929	1931	1933	1935
Belgium	0.8	6.8	10.6	11.1
Denmark	8.0	9.0	14.5	10.0
France	1.2	2.2	n.a.	n.a.
Germany‡	5.9	13.9	14.8	6.5
Ireland	n.a.	n.a.	n.a.	n.a.
Italy	1.7	4.3	5.9	n.a.
Japan	n.a.	n.a.	n.a.	n.a.
Netherlands	1.7	4.3	9.7	11.2
UK	7.2	14.8	13.9	10.8
USA	3.1	15.2	20.5	14.2

	1959–1967 (average)	1973	1977	1979	1981 (1st quarter)	1981 (December)†
Belgium	2.4	2.9	7.8	8.7	10.6	12.9
Denmark	1.4	0.7	5.8	5.3	8.0	9.5
France	0.7	1.8	4.8	6.0	7.1	8.9
Germany‡	1.2	1.0	4.0	3.4	4.1	6.7
Ireland	4.6	5.6	9.2	7.5	10.0	11.5
Italy	6.2	4.9	6.4	7.5	8.3	9.6
Japan	1.4	1.2	2.0	2.0	2.1	2.2
Netherlands	0.9	2.3	4.1	4.1	6.3	10.2
UK	1.8	2.5	5.7	5.3	9.4	11.3
USA	5.3	4.9	7.0	5.8	7.2	8.9

Sources: 1929–1935: Maddison (1980). These estimates are rather lower in many cases than those from other sources. 1959–1981: European Economy Annual Economic Report 1980-1, Commission of the European Communities, November 1980; US Department of Commerce, 1980; *Statistical Abstract of the United States from Colonial Times to the Present*, Basic Books New York, 1976; OECD 'Main Economic Indicators'.

  *1929–1935 as a percentage of total labour force; 1959–1981 as a percentage of civilian labour force.

  †Seasonally adjusted latest figures, mainly from *The Economist* 30 January 1982.

  ‡The Federal Republic for the period 1959–1981.

the figures as reflecting, in large measure, a 'socially necessary' or 'natural' state of affairs, or otherwise maintain that a proper interpretation implies that current unemployment is less unacceptable than a cursory examination of the data suggests. We shall briefly touch on such arguments below, but it is not our intention to devote a great deal of space to them. Our own position is that, despite inaccuracies in the data and despite the range of interpretations of them, unemployment now clearly represents a major social problem in most developed countries, and hence is a priority issue for analysis and policy.

Table 1.1(b) Absolute numbers of unemployed (1929-1935 , 1973-1981) (in thousands)

	1929	1931	1933	1935	1973	1977	1979	1981
Belgium	6	41	62	66	87	308	352	488
Denmark	43	53	97	76	21	147	138	214
France	10	64	305	464	576	1072	1350	1817
Germany	1899	4520	4804	2151	273	1030	876	1343
Ireland	21	25	72	123	66	106	90	132
Italy	301	734	1019	964*	1305	1382	1653	1979
Japan	n.a.	n.a.	n.a.	n.a.	680	1110	1170	1333
Netherlands	28	138	323	385	117	207	210	404
UK	1216	2630	2521	2036	575	1484	1391	2871
USA	1550	8020	12830	10610	4304	6856	5963	7900
Total (excl. Japan)	5074	16225	22033	16875	7324	12592	12023	17148

Source: Mitchell (1975), EEC (1981) for European Countries; US Department of Commerce (1980) and OECD (1981) for USA and Japan. These figures are from different sources than in Table 1.1(a); in particular the 1929–35 estimates cannot be compared with the estimated ratios of Maddison given in Table 1.1(a).
*1934

Statistics on unemployment are compiled from a count of people registered as looking for work. They are thus liable to misrepresent the true extent of the problem by the inclusion of those who, under some definitions, should not be considered as unemployed, and by the exclusion of those who, while unregistered, should be considered as unemployed. In the former category — those wrongly included — most attention has been given recently to the possibility of widespread 'voluntary' unemployment and to a suggested rise in the numbers of fraudulent claims of unemployment benefit by those actually working (in the so-called 'black economy'). The latter category — those wrongly excluded — comprises people who, while jobless and potentially interested in employment, do not register as unemployed because of pessimism regarding the likely availability of suitable job opportunities. Further factors which may cause unemployment data to give a false impression of the 'true' state of the labour market are variations in rates of part-time working and in 'labour hoarding', and temporary Government-sponsored 'job creation' and training schemes.

It is clear that this is, in quantitative terms, a very 'fuzzy' area; the extent to which such factors as those given above distort the data depends largely on arbitrary definitions and is often, in the nature of things, extremely difficult to estimate.* Most would agree, however, that the data of Tables 1.1(a) and 1.1(b) are sufficiently valid to be indicative of a highly disturbing trend.

Some further comments on these tables are appropriate. First, the authorities in different countries use somewhat different methods in compiling unemployment statistics; for example, people unemployed but temporarily sick are included in the USA and Italy, but excluded in the UK and Belgium.† The International Labour Office and the US Bureau of Labour Statistics have both attempted to compile internationally comparable series. It turns out that, for recent years at least, the changes needed are marginal in most cases, exceptions being Germany and, in particular, Italy. For 1979, for example, unemployment rates in these countries adjusted to US definitions are reduced from 3.4 per cent and 7.5 per cent to 3.0 per cent and 3.9 per cent respectively.‡ The dramatic decline in the Italian case is apparently due to the exclusion of individuals who have not actively sought work for thirty days or more. However, it is doubtful whether the revised figure portrays the Italian labour market

---

*For further discussion in the context of the UK see Garside (1980).

†The *UK Employment Gazette*, August 1980, gives details of international differences.

‡The figures and explanations for the changes are taken from Sorrentino (1981), who presents the US Bureau of Labour Statistics adjusted rates for nine countries for 1959–1979.

position more accurately than the original; under-utilization of labour in that country is reflected in very large scale involuntary part-time working (combined, so it is widely reported, with a substantial 'black' economy).

A second important point is that there are special reasons to account for some of the differences in measured unemployment rates between countries. Part of the reason for the low rate in West Germany (and also in Switzerland), for example, has been their ability to 'export' unemployment by turning the high inflow of 'guest' workers from abroad in times of labour shortage prior to 1973 into a net outflow from 1974 onwards. The number of immigrant workers in Germany declined from 2.4 million in 1973 to 1.4 million in 1977 (Maddison 1979); and, surprisingly, of the countries listed in Tables 1.1(a) and 1.1(b), Germany is the only one to have registered a *decline* in total *employment* between 1973 and 1979. In the case of Japan, the 'paternalistic' tradition of 'lifetime' employment may contribute to their success in keeping unemployment rates low, although this is probably far less important than this country's industrial and technological success in world competition. The USA is also notable as a country where until recently unemployment rates do not appear to have risen as rapidly as in Europe, and one in which new employment grew rather rapidly. We shall return in Chapter 9 and Chapter 10 to discuss some of these differences in relation to international technological competition and technology policies.

In conclusion, while there seems to have been some increase in the numbers of 'marginal' cases included in the unemployment figures, there has probably also been an increase in the numbers of those 'wrongly' excluded from the count of unemployed (i.e., people who would register if levels of recruitment improved). It is obviously difficult to estimate the net effect of these factors but both casual observation, and the movement of other indicators, suggest that the rise indicated by the unemployment statistics is certainly not a hopeless distortion of the extent to which people are involuntarily unemployed. In so far as biases are present, the evidence seems to indicate that the net effect of errors in the data is such that the official figures significantly *underestimate* the extent of unemployment in most economies.

Finally, mention should be made of important changes over time and differences between countries in the growth rate of the labour force and in the relative participation of women. As Sorrentino (1981) argues, the higher rates of unemployment in North America prior to 1973 compared with Western Europe may be attributable not only to differences in rates of economic growth, but also to a more rapid growth in the North American labour force. A comparison of the growth of the labour force in the US with that of Germany, Italy and the UK (Table 1.2) illustrates this relatively rapid growth. That much of the increase in the US labour force is accounted for by increased participation of women is shown by Table 1.2, which illustrates the dramatic increase in the female proportion of the workforce. With the exception of Japan, this proportion has increased in all the countries shown, but only since the mid to late 1960s in most cases.

The present higher rates of unemployment in Europe also may reflect in part the 'bulge' in the number of young people entering the labour market. Clearly the level of unemployment in any country reflects a complex and changing balance between *demand* for labour arising from developments in the economy and in technology and the *supply* of labour affected by demographic changes, migration flows, changes in social policy, education and attitudes. In this book there is a considerable amount of discussion on the ways in which technical change may affect *demand* for labour in different ways, at different times, and in different places, but there is no attempt to discuss the demographic, social, and other factors affecting the *supply* of labour.

The total labour force (i.e., the number of people registered as working or available for work) increased over the period 1973-1979 by more than over the period 1966-1972 in both the total OECD area and within the EEC. Total *employment* also increased over this period despite the rise in *unemployment*. It is arguable, therefore, that the recent rapid rise in unemployment resulted at least partly from the rapid growth of labour supply. However, we have chosen to focus our discussion on the forces which prevented total labour demand from keeping pace with the increased supply of labour. There is ample historical precedent for believing that there is no intrinsic reason why a rapid growth of the labour force should be accompanied by rising unemployment; Germany in the 1950s, Japan in the 1960s and the USA in the late nineteenth and early twentieth centuries offer examples of economies which were able to absorb rapid growth in the labour force and achieve sustained low levels of unemployment. In addition, a major focus of our interest is in the factors determining employment in manufacturing industry which, as Table 1.2 shows, has declined in most countries since the 1960s.

Table 1.2 Labour force,* manufacturing employment (in millions) and female participation, various years

		1954	1958	1962	1966	1970	1974	1978	1980
France	Labour force	19.6	18.8	19.8	20.6	21.4	22.2	22.9	23.2
	% female	34.4	34.6	33.4	n.a.	35.0	36.2	37.8	38.3†
	Manufacturing employment	5.0	5.3	5.3	5.6	5.7	5.9	5.6	5.4
Germany	Labour force	n.a.	26.3	26.9	27.3	26.8	26.8	26.2	26.7
	% female	n.a.	37.2	26.9	36.1	35.9	36.9	37.7	38.0
	Manufacturing employment	n.a.	8.4	9.3	9.5	9.8	9.4	8.6	8.7
Italy	Labour force	20.6	21.9	20.8	19.8	20.9	21.1	22.1	22.8
	% female	30.4	31.3	29.5	26.4	28.8	26.7	32.1	33.3
	Manufacturing employment	4.2	5.1	5.6	5.4	5.9	6.1	n.a.	n.a.
Japan	Labour force	40.6	48.9	46.1	48.9	51.5	53.1	56.0	56.5
	% female	40.7	41.1	40.3	39.9	39.3	37.7	38.0	38.7
	Manufacturing employment	7.4	9.0	10.7	11.8	13.8	14.3	12.8	12.8
UK	Labour force	24.3	24.7	25.6	26.2	25.3	25.6	26.2	26.4
	% female	32.3	32.9	24.0	35.0	35.3	37.5	38.7	39.1
	Manufacturing employment	8.1	8.2	8.6	8.7	8.5	8.0	7.3	6.9
USA	Labour force	67.8	70.3	73.5	78.9	85.9	93.2	102.5	106.8
	% female	29.4	31.5	32.7	34.6	36.7	38.5	41.0	41.9
	Manufacturing employment	21.6	21.2	22.2	25.0	25.4	26.7	27.5	27.3

Sources: OECD Labour Force Statistics, Paris, various years; Brown and Sheriff (1979).
*'Labour Force' = total employed plus total registered unemployed.
†1979.

1.2 A return to 'full employment'?

It is clear from the inclusion of certain categories of persons in the unemployment statistics that the measured rate of unemployment cannot fall to zero. This raises the question of whether it is possible to determine a level which can for practical purposes be regarded as 'full employment', and which can therefore form a realistic policy goal.

The best known numerical definition of full employment is that due to Lord Beveridge (1944) who adopted the figure of 3 per cent unemployment or below, at the seasonal peak, as signifying a state of full employment. This figure is, of course, to a great extent arbitrary; as Tables 1.1(a) and 1.1(b) indicate, the average levels for the period 1959-1967 were significantly below this on average for many countries, and other data show that it is not unprecedented for a figure below 2 per cent to be sustained for many years. More recently, however, a 3 per cent goal would be considered as unrealistically low by many observers; the level of unemployment corresponding to 'full employment' in 1982 would widely be regarded as higher than it was in 1962. Some discussion of the notion of 'full employment' and its relation to changing circumstances is therefore in order.

In the 1950s, Rees (1957) gave a comprehensive discussion of possible measures of full employment; five broad possibilities are put forward, some of them being subdivided into further alternatives. Three of these five categories — the minimum unemployment approach, the maximum employment approach, and the turnover approach — are based on historical experience. The 'minimum unemployment approach', for example, defines full employment as being present when unemployment is equal to the lowest value previously achieved. Such a definition, while possibly useful as a starting point, suffers from the defect that, because of changing circumstances, the 'irreducible minimum' of an earlier period may differ substantially from that of the present. There is also the implication that the earlier minimum is not only achievable but sustainable (i.e., that it did not, even in its time, represent a situation of 'overfull' employment). A fourth method discussed by Rees is the 'unfilled vacancies approach' which he compares with Beveridge's definition that full employment 'means having always more vacant jobs than unemployed men'. This measure is free from the problems associated with historical analogy but suffers from the difficulty of quantification. Reliable estimates of unfilled vacancies are difficult to obtain and, even if they are available, it is hard to

see how a figure for the 'full employment' rate of unemployment
can be derived from them. Data on vacancies do, however, provide
some clues as to the nature of measured unemployment, and to
differences between countries, a point to which we will return
later in this chapter.

The fifth measure discussed by Rees, the 'price approach', is
closely related to the approach used in modern monetarist theory.
The notion here is that, if unemployment declines below a certain
level, the result is necessarily an increase in prices. At least in
principle, 'full employment' can then be defined as consistent with
that level of unemployment corresponding to stable prices, a level
which is likely to be higher than the recorded historical minimum.
With such a definition, the inflationary pressures of the 1970s,
combined with high unemployment levels, suggest that the 'full
employment level of unemployment' appropriate to the 1970s
and early 1980s is dramatically higher than that of the previous
two decades and hence that the extent to which high unemployment could, or should, be subject to macroeconomic policy is
circumscribed.

Monetarist theory takes this argument further. Full employment
is defined in terms of the 'natural' rate of unemployment, which
is often taken as the rate at which inflation is correctly anticipated
(but is not necessarily zero). Essentially, the natural rate corresponds
to an equilibrium state in the labour market where, at the existing
real wage, the number of people offering themselves for work equals
the number of jobs offered by employers, and if this state of affairs
actually exists then the rate of inflation remains unchanged. The
price of pushing unemployment below the natural rate is an increase
in the rate of inflation; this will continue to accelerate unless, and
until, unemployment returns to the natural rate.

The theoretical basis of the different definitions of full employment diverge most strongly in respect of their assumptions regarding
the effects of aggregate demand and wage inflation. In general,
definitions of the 'historical minimum' type are implicitly or explicitly based on the idea that the state of demand is the variable
most critical in determining how closely 'full employment' is
approached. 'Demand-deficient' unemployment is of course regarded
by neo-Keynesians as directly amenable to correction by macro-
economic policy; such thinking formed the conventional wisdom
up to the early 1970s, and while the events of more recent years
have given rise to considerable divergence of viewpoints, it is probably true to say that demand deficiency is still regarded as a major,
if not the major, cause of unemployment by most commentators

and national governments. Many Keynesians regard wage inflation as by and large a 'cost-push' phenomenon, wage and/or price controls frequently being advocated as anti-inflationary measures. Unemployment resulting from demand deficiency is sometimes called 'cyclical' owing to observed fluctuations of demand over the short-term business cycle, although the possibility of persistent demand deficiency over a period of a decade or more cannot be ruled out, particularly if Government policy is deflationary.

The monetarist view ascribes little or no importance to demand deficiency, at least in a period of constant or rising inflation. Aside from frictional and structural effects, full employment can be achieved by the interplay of supply and demand in the labour market. Rising unemployment may be due to a rise in the (full employment) 'natural rate' which according to some could arise partly from an increase in 'voluntary' unemployment. We share Kristensen's (1981) doubts about the use of the expression 'natural' in this context (see Chapter 10) but a fuller discussion of the natural rate hypothesis is inappropriate here; it is in any case discussed in detail in many recent texts (e.g., Chrystal 1979, Trevithick 1980).

All definitions of full employment recognise the existence of unemployment which is sometimes called 'frictional' — for example, people moving between jobs — and which, by definition, cannot be entirely eliminated, although it may be reduced by improving vacancy information services. There is also agreement on the existence of some degree of so-called 'structural' unemployment: the distinction between frictional and structural unemployment is not sharp, although by implication the latter reflects some kind of imbalance between the structure of the workforce (in terms of, for example, age and skill profiles and geographical location) and the particular mix of worker attributes required by the sectoral and technological characteristics of the production system. Such mismatches can arise in the general process of economic change and may be mitigated by such measures as relocation, retraining or increasing or changing the nature of the productive potential of the economy through fixed investment. On the other hand, economic change may be such that structural problems become progressively more serious, perhaps because of self-reinforcing trends in the economy. This possibility will receive considerable attention in the course of this book.

In purely 'static' terms, it may be asserted that involuntary unemployment must arise either because total demand in the economy is not sufficient to ensure that the quantity of goods and services that can be provided at full employment are purchased; or, on the supply side, because of a constraint in some factor of production. A 'snapshot' of an economy may reveal a high degree of under-utilization of capital in some sectors, implying demand deficiency, combined with a shortage of particular skills or of productive capacity in other industries where domestic or overseas demand may be readily forthcoming. These problems — reflecting, respectively, inadequacies in the *demand* and *supply* of goods — are therefore by no means mutually exclusive, and neither are they independent: demand may be created by supply (for example of a new product), and *vice versa*. It is, indeed, a major focus of our later arguments that it is the changing dynamics of the interactions between demand and supply which is central to explaining changing economic fortunes. Bearing these points in mind, however, it is useful to summarise the conditions that can accompany unemployment and outline some relationships of technical change to them, as these are relevant to some of our later discussions.

The possible reasons for the demand for labour being insufficient for 'full employment' include:

(i) *Persistent insufficient aggregate demand for goods and services*. This will occur if the main categories of demand-consumer expenditure, investment, exports and Government expenditure are, in sum, too low to correspond to full employment, in other words if:
(a) consumers try to save a high proportion of their income;
(b) private investment is low;
(c) export demand is slack;
(d) government is unwilling to intervene to the required extent for political reasons or through fear of inflationary effects or external balance problems.
Causes (a), (b) and (c) may be exacerbated by a relative dearth of new technologies, and are related in that a perception of declining consumer demand propensities by entrepreneurs will adversely affect their incentive to invest. The relationship of technical change to (d) depends on one's view of the causes of, and cures for, inflation (Cooper and Clark, 1982); rapid productivity growth may tend to ease the inflationary constraint and there is evidence that product and process innovation is an important factor in international competitiveness and bears heavily on the balance of trade issue (Pavitt and Soete, 1980).
(ii) *Mismatches in the labour force*. These may take the form of a widespread lack of appropriate skills, due to rapid technical change or to changing tastes, leading to changes in industrial structure; and/or the form of geographical separation of job opportunities

and workers. Depending on the technologies available, a skill shortage may be 'absolute' in the sense of being independent of relative prices or wages; or be dependent on institutional inflexibilities in wage and price movements which prevent factor substitution (e.g., substitution of capital for skilled labour) which in principle could ensure full employment.

(iii) *Shortages or mismatches in productive capacity*. Precisely analogous arguments to those regarding skills apply to the stock of fixed capital. An absolute shortage may be said to exist if there has been a history of insufficient investment activity, and unemployment remains even if all fixed capital that can be operated profitably at arbitrarily low wage levels is fully utilized. More realistically, inflexibilities in wages and in the related response of relative prices to changing demand patterns and technical change could render a proportion of the stock of available fixed capital economically non-viable, the remainder (i.e., that which in practice can be operated profitably) being inadequate to provide full employment. This situation may be looked upon as essentially equivalent to the familiar argument that unemployment is caused by wages being too high (i.e., by workers 'pricing themselves out of jobs'). Whether the 'cause' is regarded as inadequate entrepreneurial activity or unrealistic wage claims is, in large measure, a matter of political predisposition.

Any attempt to allocate measured unemployment between these categories is extremely hazardous. The most practical assumption is to regard unemployment existing at peaks in the business cycle as being related to factors other than demand deficiency (i.e., to 'structural' factors (ii) and (iii) above) combined with frictional and voluntary components. However, it is not implausible that the rise in unemployment over the past decade has been accompanied by a persistent deficiency of demand, and hence that recent 'cyclical peaks' are not strictly comparable to those of earlier periods. Some indication of the state of the labour market may be obtained from an examination of job vacancy statistics. Table 1.3 presents estimates of the ratio of job vacancies to unemployment in six countries. In France and Belgium, as well as in the UK, the number of vacancies has fallen below 10 per cent of the number of unemployed. In West Germany, the ratio has fallen dramatically since 1973, despite a partial recovery after 1976. In Japan the improvement following the mid-1970s recession is stronger, while for the USA the more recent values do not appear to be significantly different from those of the early 1970s. The tentative conclusion is that those countries where increases in unemployment have been most pronounced are

Table 1.3 Vacancies/unemployment ratios

	1963	1972	1973	1974	1975	1976	1977	1978	1979	1980
Belgium	0.30	0.10	0.15	0.13	0.02	0.02	0.01	0.02	0.02	0.02
France	n.a.	0.44	0.64	0.43	0.13	0.13	0.10	0.08	0.07	0.06
West Germany	3.1	2.2	2.0	0.54	0.22	0.22	0.23	0.25	0.35	0.34
Japan	0.62	0.90	1.31	0.82	0.39	0.47	0.50	0.47	0.59	0.61
UK	0.38	0.23	0.68	n.a.	0.17	n.a.	n.a.	0.15	0.19	0.09
USA	n.a.	0.62	0.84	0.63	0.30	0.39	0.52	0.74	0.79	0.52

Source: OECD, 'Main Economic Indicators', various years.

characterized by a *lack* of vacancies rather than a substantial pool of 'inappropriate' vacancies.

Data presented by Scott (1978) for the UK show that, for *male* employees, a given level of vacancies has, since 1966, corresponded to a higher and increasing level of unemployment, suggesting an increasing trend in that proportion of unemployment *not* related to structural problems in the labour force. For female employees there is a suggestion of a similar trend prior to 1975, although it is much less marked; after 1975, however, there is an extremely rapid increase in unemployment with no clear trend in vacancies. Social and institutional factors, together with changing attitudes, may help to explain these important differences.

While again pointing out the interdependence of the various factors, we are inclined to stress the importance of aggregate demand deficiency in the present context, combined with problems associated with the magnitude and structure of productive potential in several countries. In Chapter 2 the nature of inflexibilities affecting the growth and structure of the capital stock are discussed and in Chapter 8 some empirical evidence will be presented to suggest that supply-side difficulties resulting from inflexibilities in the stock of fixed capital may be experienced in the event of an upswing in demand.

1.3 Technical change and unemployment

Part of the difference in emphasis between economists on such issues as structural unemployment, or whether 'technological' unemployment ever really exists, lies in their degree of faith in the efficacy and importance of the self-adjusting market mechanisms. In truly perfect markets even frictional and short-term cyclical unemployment would disappear, and neither technical nor demographic change would present any problems. It is universally recognized that, in practice, markets are far from perfect and that adjustment processes, which might ultimately match demand and supply, do so only with time lags which are sometimes prolonged. However, those with greater confidence and faith in the efficacy of market mechanisms tend to regard both technical change and demographic change as either rather minor, temporary disturbances or as continuous processes which need not and do not deflect the economy from an otherwise smooth equilibrium growth path. Technical change, for example, is sometimes believed to generate compensating effects which will almost automatically compensate for any labour displacement by new demands for employment elsewhere in the economy. In some other formulations, technical change is itself part of the adjustment process.

Other economists, however, have had far less faith in these equilibrating tendencies. Those economists who have been occupied with the problem of long-term unemployment, such as Marx and Keynes, have stressed particularly fluctuations in investment behaviour and lags in the capital accumulation processes. Pasinetti (1981) and Cornwall (1977) are recent examples of economists who have pointed to the problems of prolonged unemployment associated with changing patterns of investment and growth, inflexibilities in prices and in substitution processes. On any showing investment represents a critical and volatile component of aggregate demand, and is therefore central to outstanding 'demand deficiency' problems; while in the longer term, it leads to an increase in productive capacity which, following many years of depressed net investment activity, seems likely to have become a constraint on a return to full employment. We shall later also discuss the possibility that a change in the nature (as well as the volume) of investment activity is of importance to employment.

The effects of technical change on the level of employment depend not only on the kind of technical change involved but also on the overall economic situation. If large gains in labour productivity become feasible through technological advance, the implications for employment depend on the extent to which such gains are actually achieved and how they compare with the growth in output during the period in which the technological diffusion process is occurring. This output growth may result partly from the availability of the technology itself but may also be totally independent of it. Given the complexity of the issues involved it is not surprising that a range of opinions are possible concerning the employment impacts of new technologies, as recent debates related to microelectronics have shown. The conflicting viewpoints result chiefly from differing assumptions (implicit or explicit) regarding the following factors:

(a) the likely rate of diffusion of the technology, both domestically and relative to other (competitor) countries;
(b) the relative influence on direct demand creation and labour displacement;
(c) in the case of a perceived direct reduction in labour demand, the possibility of compensation through price feedback or through Government intervention;
(d) the extent to which new vintages of capital equipment may require less labour and how flexible such technical coefficients may be.

As mentioned above, the concept of 'compensation' implies that mechanisms exist whereby the tendency for workers to be displaced by labour-displacing technical change is associated with adjustment processes which lead to such workers being reabsorbed, either in the same industry or elsewhere — for example, in the capital goods industries making the new machinery. Alternative theories of compensation have been extensively discussed by Gourvitch (1940), Neisser (1942), and, more recently, by Heertje (1977) and Blattner (1979). The more recent analyses of the process involved centre around the idea of neoclassical price adjustment: for an arbitrary number of products and factors of production (fixed capital, labour etc.), and given any set of fixed technical coefficients (i.e., for any set of technologies) it is in general possible to find a set of product and factor prices which ensures the full use of resources (i.e., full employment). It is, however, quite likely that meaningless results will be obtained (e.g., wages or some product prices may be negative). Neisser (1942) gives an example of this. Such results occur if there is an absolute shortage of one factor (e.g., skilled labour or certain types of fixed capital), such that *no* pattern of final demand for products ensures full employment of a relatively abundant factor (e.g., unskilled labour). This situation may be described as true 'technological unemployment' — with the technological system and available resources given, no movement of prices or wages will ensure full employment. Such a situation may be mitigated by relaxing the condition of fixed technical coefficients to allow substitutability between factors of production, to an extent determined by the degree of such substitutability.

In the real world, however, factor rewards can be expected to have (positive) lower limits, or be 'sticky downwards'. Thus it may be possible for firms in a monopolistic or oligopolistic situation to fix a certain rate of return on capital, or union pressure may ensure that wages are inflexible downwards and/or that relative wages remain immutable. Thus even with alternative methods of production allowing for substitutability between factors, institutional factors such as those mentioned would inhibit such substitution under technical change. In this case we may speak of the possibility of 'technologically induced' unemployment which *results* from technical change but is dependent on institutional constraints to adaptation (see Cooper and Clark (1982) for a fuller discussion).

The difficulty of identifying and quantifying feedbacks and problems in estimating time lags are major reasons for differences of view over the employment effects of new technologies. However, the notion that technical change is something to which 'adjustment'

is required, the net employment effects of the change being determined by the extent to which such adjustment takes place, is, in
any case, a big over-simplification and a rather limited approach
in most respects. The general position taken in this book is that —
because the impact of technology on the economic system is so
pervasive, and yet so uneven in time and space — the employment
effects of technical change may be more fruitfully analysed in
terms of the dynamic mechanisms that may be set in motion by
technical change rather than with reference to notional 'equilibrium'
or 'static' situations. In the following chapter we introduce the idea
of 'long waves' in economic development and argue for the importance of technical change both as a stimulus to an upswing in investment activity — an 'engine of growth', and as a response to the
changing pattern of demand and prices — a 'thermostat' (Dosi, 1982).

2 SCHUMPETER'S THEORY OF BUSINESS CYCLES AND INNOVATION

As we have shown in Chapter 1, unemployment in the early 1980s in most of the industrialized countries has been higher than at any time since the 1930s. These changes were not merely the result of short-term fluctuations but reflected a secular trend associated with generally slower rates of growth, slackening investment and more determined efforts to contain the growth of government expenditure. It is now increasingly recognised that the period of very rapid economic growth which succeeded World War II has given way to a rather different phase of recession and 'stagflation' — slower growth associated with much higher levels of unemployment. In these circumstances it is hardly surprising that interest has revived in theories of Kondratiev 'long cycles' or 'long waves' in economic development, which seek to explain such long-term changes in the economic climate. In this chapter we are concerned in particular with the explanation of long cycles advanced by Joseph Schumpeter, who, more than any other twentieth-century economist, attempted to explain competition and economic growth largely in terms of technical innovation. An interesting personal comment on the contemporary relevance of these ideas came from an outstanding Keynesian economist of the post-war period — Paul Samuelson (1981):

No one can predict the future with confidence. Still, it is my considered guess that the final quarter of the twentieth century will fall far short of the third quarter in its achieved rate of economic progress.
The dark horoscope of my old teacher Joseph Schumpeter may have particular relevance here. When I was a precocious student, I didn't think much of Joseph Schumpeter's futurology. But, like Mark Twain, who said that when he was fourteen he thought his father was awful dumb, but by the time of reaching twenty-one he was surprised at how much the old boy had smartened up, when I reread Schumpeter's 1942 *Capitalism, Socialism and Democracy* I find new meanings in it.

There *do* appear to have been in the past century or so several periods of rather deep crises and slower growth, associated with higher levels of unemployment, which at least in some countries were regarded at the time and by historians since as 'Great Depressions'. These alternated with the periods of boom and prosperity experienced in the 1850s and 1860s, in the *Belle Epoque* before

World War I and in the 1950s and 1960s. The depressed periods were roughly the 1880s and the 1930s, although varying a little from country to country (see Chapter 9). In the USA, for example, the Civil War in the 1860s was followed by a period of rather rapid growth in the 1870s and 1880s when European countries were generally experiencing slower growth rates and more depressed conditions. Japan was less affected by the depressions of the 1930s and (so far) the 1980s.

2.1 Schumpeter's theory of long cycles

In his major work on *Business Cycles*, Schumpeter (1939) both accepted the reality of the phenomenon of 'Kondratiev'* long cycles, lasting half a century or so, and offered a novel explanation of them, differing from that of Kondratiev (1925) himself. According to Schumpeter (1939, Chapter 2), each business cycle was unique because of the variety of technical innovations as well as the variety of exogenous events such as wars, gold discoveries or harvest failures. But despite his insistence on the specific features of each fluctuation and perturbation, he believed that the task of economic theory was to go beyond a mere catalogue of accidental events, and analyse those features of the system's behaviour which could generate fluctuations irrespective of their specific and variable form. The most important of such features in his view was innovation, which despite its great specific variety, he saw as the main engine of capitalist growth and the source of entrepreneurial profit.

The ability and initiative of entrepreneurs (who might or might not themselves be inventors but more usually would not be) created new opportunities for profits, which in turn attracted a 'swarm' of imitators and improvers to exploit the new opening with a wave of new investment, generating boom conditions. The competitive processes set in motion by this 'swarming' then gradually eroded the margins of innovative profits (as in Marx's model), but before the system could settle into an equilibrium condition the whole process would start again through the destabilizing effects of a new wave of innovations. This process was sufficient in itself to

*As has often been pointed out, Kondratiev was by no means the originator of the long cycle theory and it is in some respects a misnomer that the phenomenon bears his name. The Dutch Marxist van Gelderen (1913) could be much more fairly credited with the idea, which he articulated clearly in 1913. At about the same time a variety of economists, including Pareto (1913), had drawn attention to the apparent tendency for long-term price movements, interest rates and trade fluctuations to follow a cyclical movement lasting about half a century. However, during the 1920s whilst heading the Institute of Economic Research in Moscow, Kondratiev did more to propagate and elaborate the idea than any other economist.

engender various types of cyclical behaviour, although Schumpeter certainly acknowledged that there was a process of interaction with many other features of the economic system which have been the subject of more conventional business cycle analysis.

Whether or not such a mechanism offers a plausible explanation of 'long' (Kondratiev) cycles in economic development depends crucially — as Kuznets (1940) pointed out in his review of *Business Cycles* at the time — on whether some innovations are so large and so discontinuous in their impact as to cause prolonged perturbations or whether they are bunched together in some way. The construction of a national railway network might be the type of innovative investment which would qualify as a 'wave generator' in its own right, but obviously there are thousands of minor inventions and technical changes which are occurring every year in many industries whose effect is far more gradual and which might well adapt to some sort of smooth equilibrium growth path. If these smaller innovations were to be associated with economic fluctuations then this could only be if they were linked to the growth cycles of new industries and technologies:

A neo-Schumpeterian interpretation of the post-war boom would see it primarily as the simultaneous explosive growth of several major new technologies and industries, particularly electronics, synthetic materials, drugs, oil and petro-chemicals, and (especially in Europe and Japan) consumer durables and vehicles. It can be shown that these fast-growing industries and the associated component and machinery suppliers accounted for a large part of the growth of industry in many OECD countries (Chapter 7). It is interesting to note that neither Keynes nor most Keynesians expected that the quarter century after the war would be the fastest period of economic growth the world had ever experienced. Indeed the expectations of economists immediately after World War II were mostly pessimistic. This was partly because they tended simply to extrapolate the problems of the 1920s and 1930s into the post-war world and made no allowance for new technologies and the impetus they would give to profit expectations and to new investment. When, however, the big boom did materialize the tendency was to attribute its success to the adoption of Keynesian policies. An important exception was Robin Matthews (1968), who already in 1968 in an article in the *Economic Journal* on why Britain had full employment since the war, pointed out that it was the byouancy of private investment rather than government policies which sustained the boom. Other economists (Dow 1964) pointed out that Keynesian policies were often inept since they were applied after the relevant turning point in the business cycle had already occurred 'spontaneously'.

Schumpeter had suggested that after a strong 'band-wagon' effect and the entry of many new firms into the rapidly expanding sectors attracted by the exceptionally high profits of innovation, there would follow a period of 'competing away' of profits as the new industries matured. This could lead to stagnation and depression if a new wave of innovations did not compensate. Such an explanation appears to fit the facts of the post-war boom. It had been remarked already in the 1960s that the general rate of profit was beginning to fall in several OECD countries and this tendency was aggravated by more severe international competition (Chapter 9). This became still more marked in the 1970s especially in some of the erstwhile rapid growth sectors such as synthetic materials and consumer durables. In spite of its continued high rate of growth, the electronics sector was also affected by this general trend (Chapter 6).

Schumpeter had relatively little to say about unemployment and wages, but economists of many different schools, including both neo-classical and Marxist, would agree that a sustained period of boom and full employment such as that experienced before World War I or after World War II would tend to strengthen the bargaining position of labour and thereby also to erode the rate of profit and stimulate cost-push inflation, with or without the pressure of militant trade unions. These changes would tend to induce rather different types of technical change, and associated investment. In the early period of a long boom the emphasis is on rapid expansion of new capacity in order to get a good market share and this investment has a strong positive effect on the generation of new employment. As the new industries and technologies mature, economies of scale are exploited and the pressures shift to cost-saving innovations in process technologies. Capital-intensity increases and employment growth slows down or even stops altogether. Again, this hypothesis appears to fit the post-war pattern quite well. Already before the OPEC crisis these trends were apparent in the leading industrial countries, both in the newer industries such as synthetics and electronics, and in older ones.

We shall attempt to illustrate these points for synthetic materials and electronics in Chapters 5 and 6, and for the economy more generally in Chapters 7 and 8. But first we shall attempt to answer two questions: why was Schumpeter's theory rejected or neglected by most economists until recently? And how does his theory relate to more traditional explanations of business cycles including Keynesian theory?

2.2 The debate on Kondratiev and Schumpeter

One reason for the neglect of Schumpeter was the general reluctance of the economics profession to tackle the thorny problems of invention and innovation, which has already been alluded to in the introduction. Another reason was that Schumpeter's book on *Business Cycles* appeared in 1939 (i.e., three years after publication of Keynes' (1936) *General Theory*, which by then and for a long time afterwards occupied the centre of the stage in the professional debates on cycle theory and policy-making). As even his warmest admirers would agree, *Business Cycles* is a badly written book. It is inordinately long and the use of statistics is poor. Fels (1964) performed a valuable service in producing a much more intelligible abridged edition, but this was quite a while later and even this edition is not an easy read. Perhaps most important of all Schumpeter did not satisfactorily explain why major innovations or clusters of innovations should occur only every half century or so. Indeed the whole notion of long waves or cycles was rather discredited in the 1950s and 1960s, both in the West and in the East.

Kondratiev himself was one of Stalin's victims in Siberia but before this he had come in for thorough criticism on the part of his orthodox Marxist opponents on the grounds of his supposedly faulty statistical techniques, problems in the dating of turning points of the various cycles, inaccuracies and discrepancies in his data, paying too much attention to prices and interest rates rather than production and lack of a satisfactory theoretical explanation of long cycles consistent with Marxist theory (see Barr 1979). Some of these criticisms were unjustified and prejudiced. The notion that capitalism might enjoy a renewed upsurge of growth after the depression was particularly unwelcome in the 1930s. However once the ideas became known in the West, they were subjected to criticism almost as severe as in the East, notably by Garvy (1943) and Weinstock (1964). Although the notion of long waves had its defenders, both among monetarists such as Dupriez (1947) and among Marxists such as Mandel (1972 and 1980), for the most part economists either rejected or ignored them.

During the 1970s, however, much has changed, justifying Mandel's (1981) ironic comment that it takes the prolonged down-turn of a long wave to compel many economists to reconsider their ideas and to open their minds to new conceptions. There has been a notable quickening of interest in long waves in the last few years. This has been most strongly marked in the Netherlands and Belgium where the ideas always had a wider currency, but it is also apparent in Germany and France and the USA and there have even been some signs of a revival of interest in Eastern Europe. The research programme at Amsterdam University led by van Roon* is one indication of this revival and another was the conference on long waves at Bochum University in 1980 (Petzina and van Roon 1981).

At this conference there was a lot of discussion on the statistical aspects of long wave theory. This is not surprising because among the other main reasons for the rejection of the long cycle or long waves idea, whether in its Schumpeterian or any other form, were and are the statistical problems associated with any attempt to reconstruct the economic time series of the past 200 years, or even the last 100 years. Even if this can be done, there are big problems over the interpretation of such series in relation to the incidence of major events such as the two world wars or the US Civil War. There are other problems of a purely statistical nature as well as the major difficulty of building any generalizations on only four cycles extending over a period which witnessed enormous social and political changes, and in which only a few countries were involved in the early cycles. Finally many people are uncomfortable with the implicit or explicit mechanistic determinism of some long wave theories and in some cases for their assumption of a fixed periodicity. For these reasons, the work of the three Kiel economists, Glismann, Rodemer and Wolter (1980) is particularly interesting. As in our case, the focus in their analysis is on long-term *production* and *investment* fluctuations, rather than monetary and price fluctuations. Their analysis contains fairly good statistical evidence for the past century on long-term cycles for various countries.

In Figures 2.1(a) and 2.1(b) we have replotted some of the statistical evidence collected by Glismann, Rodemer and Wolter, not by individual country, but rather for the total of the various countries included in their analysis: Britain, France, Germany, Italy, Sweden and the USA for production, and Britain, Germany, Italy, Sweden and the USA for investment,† these being assumed to approximate total world

---

*Further information available from Prof. G. van Roon, Coordinating Director of the Research Group on Long-term Fluctuations at the Free University of Amsterdam.

†Only for Britain do Glismann, Rodemer and Wolter provide data from 1830 onwards. To avoid the abrupt addition of the various other countries' production and investment figures (which began in 1850 for Germany, in 1861 for Italy, in 1861 for Sweden, in 1889 for the USA and in 1900 for France) to only the British figures, we estimated production and investment for each country from 1830 onwards, by extrapolating backwards the exponential trend through the production and investment data of each country from the first available year to 1913. For the years 1914 to 1924 and 1920, the missing German and French production and investment figures were assumed to be identical to the 1924 and 1920 figures. For 1942 to 1948 the missing German figures were assumed to be identical to the 1948 figures. All production and investment figures are in real 1970 prices and exchange rates.

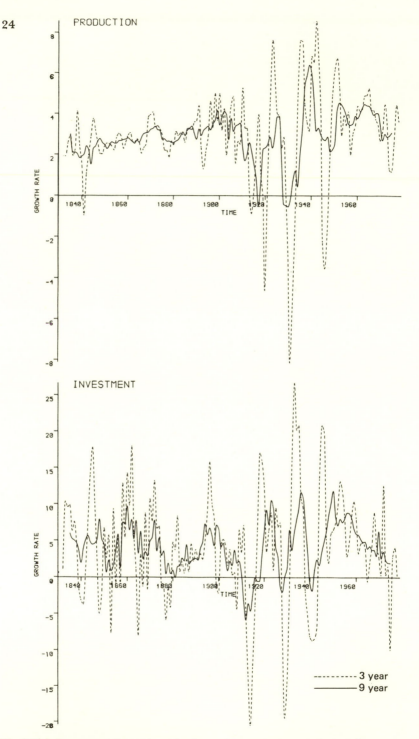

Fig. 2.1(a) Long waves in total production and investment: growth rates (three- and nine-year moving averages)

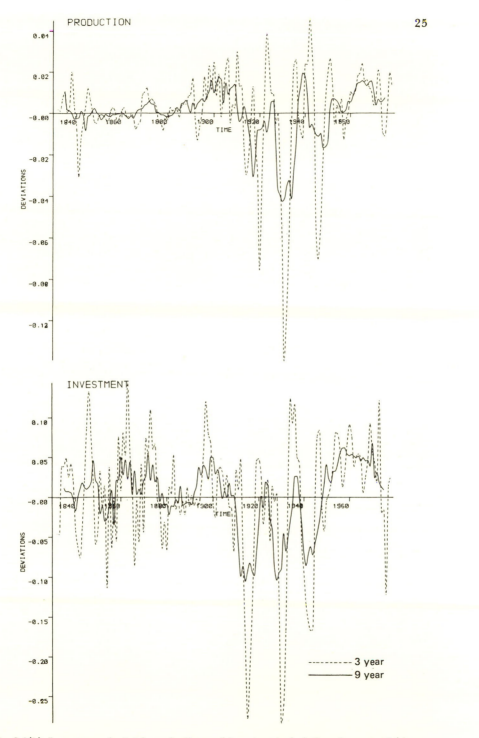

Fig. 2.1(b) Long waves in total production and investment: deviations from trend (three- and nine-year moving averages)

production and investment. As will be argued in Chapter 9, the phenomenon of long waves is in the first instance an overall 'world' phenomenon. Individual countries' pattern of output and investment growth will primarily be influenced by the general process of 'catching-up', which seems to have characterized most of the countries analysed in the Kiel study following the industrial revolution in Britain and the ensuing industrialization elsewhere.

Figure 2.1(a) presents the three- and nine-year moving averages of the production and investment growth rates for the *total* sample of countries for the period 1830 to 1979, while Figure 2.1(b) presents the three- and nine-year moving averages of the deviations from the growth trend for the same sample of countries and for the same period. Whether Figures 2.1(a) and 2.1(b) provide support for a 50-year long-term cycle is largely an academic question. As emphasized before, the evidence available — 150 years — is totally insufficient, while external factors — in particular the World Wars I and II — provide major disturbances. There is consequently little doubt that the statistical debate will continue for a long time.

However, to appreciate the relevance and importance of Schumpeter's ideas on technical innovations and economic fluctuations, it is by no means necessary to accept the idea of cycles as such and certainly not the notion of fixed periodicity. Van der Zwan (1979) prefers the notion of periodic major structural crises of adjustment and Mensch (1975) speaks of a 'metamorphosis' model. Almost all the advocates of the idea now prefer to speak of 'long waves' rather than 'long cycles'. For the purposes of this book it is necessary only to accept van der Zwan's concept that there have been periodic major structural crises of adjustment varying a little in their severity and timing between countries and followed by fairly prolonged periods of expansion and prosperity. These major crises were generally perceived at the time as being more severe than the ordinary down-turns of the short- and medium-term (Juglar) business cycles. The change in the economic climate during the 1970s and early 1980s obviously raises the question of whether there are any analogies or useful comparisons with the 1930s, and in particular whether technical innovation has any bearing on these issues. Among the important recent attempts to tackle these questions are those of Mensch (1975) and van Duijn (1979).

2.3 Capital accumulation, investment and employment

However, before discussing their work in Chapter 3, it is essential to place Schumpeter's theory in the context of more 'traditional' mainstream economic analysis, particularly investment, employment and growth theory. Neither Kondratiev nor Schumpeter explicitly discussed the possibility that a particular wave of innovation might first of all have big net employment generating effects but at a later stage some employment displacing effects. They were apparently thinking mainly in terms of the general climate of investment and accumulation and fluctuations in this overall level of investment. Thus the upswings of the Kondratiev would be periods in which investment was generally buoyant because of the new profitable opportunities which were perceived. The higher levels of unemployment experienced during the Kondratiev down-swings would be due to the loss of this impetus and to general demand deficiency associated with low levels of private investment and profitability and not necessarily associated with any change in the rate of job generation and displacement associated with a particular volume of investment. Central to these long-wave explanations is thus the concept of capital accumulation. In general, if most economists accept these days the possibility of some cyclical trend in economic life, it is primarily related to the role of capital accumulation (investment) in economic growth. The so-called 'multiplier–accelerator' models (Samuelson 1939) with their induced investment function, are a typical example of cyclical behaviour where delays in multiplier and accelerator effects can be prime causal factors in the cyclical over- and under-expansion of the capital-producing sector.

There are a large number of imperfections, lags and discontinuities involved in any expansion of the capital stock. These imperfections relate both to the so-called short run 'investment demand', and to long run capital accumulation.

Short-run investment demand is generally assumed to be a function of so-called 'expectations' and interest rates. If investment demand is highly interest-elastic (i.e., responsive rapidly to small changes in the rate of interest given a certain level of expectations), then depending on wage and price flexibilities, a full employment general equilibrium could easily be obtained. Recessions and unemployment could be overcome by simple monetary policies. However, little evidence exists to support this view of the interest-elasticity of investment demand. Rather, there exists a variety of views on the determinants of investment demand. These range from high optimism that a self-stabilizing full-employment level of investment demand will come forth with the lightest hand being kept on the money supply, to great pessimism about the unpredictability and volatility of investment demand, which leads to a strong preference for fiscal policy (Cooper and Clark 1982). In terms of these volatile expectations, it is always worth recalling the well-known passage from Keynes (1936, pp. 161–2):

Most, probably, of our decisions to do something positive the full consequences of which will be drawn out over many days to come can only be taken as a result of animal spirits — of a spontaneous urge to action rather than inaction, and not as the outcome of a weighted average of quantitative benefits multiplied by quantitative probabilities. Enterprise only pretends to itself to be mainly actuated by the statements in its own prospectus, however candid and sincere. Only a little more than an expedition to the South Pole, is it based on an exact calculation of benefits to come. Thus if the animal spirits are dimmed and the spontaneous optimism falters, leaving us to depend on nothing but a mathematical expectation, enterprise will fade and die; — though fears of loss may have a basis no more reasonable than hopes of profit had before. . . . This means, unfortunately, not only that slumps and depressions are exaggerated in degree, but that economic prosperity is excessively dependent on a political and social atmosphere which is congenial to the average business man. . . . In estimating the prospects of investment, we must have regard, therefore, to the nerves and hysteria and even the digestions and reactions to the weather of those upon whose spontaneous activity it largely depends.

Keynes' deep (if occasionally ironical) awareness of the subtleties of investment behaviour was based on a life-time of direct experience in the Stock Exchange and the commodity markets. It is in sharp contrast not just with the simplistic dogmatism of some contemporary monetarist prescriptions, but also with that of some of his own followers (which is exactly what he would have expected). Summing up his discussion on long-term expectations he says (Keynes 1936, p. 164):

Only experience, however, can show how far management of the rate of interest is capable of continuously stimulating the appropriate volume of investment. For my own part I am now somewhat sceptical of the success of a merely monetary policy directed towards influencing the rate of interest. I expect to see the state, which is in a position to calculate the marginal efficiency of capital goods on long views and on the basis of the general social advantage, taking an ever greater responsibility for directly organising investment; since it seems likely that the fluctuations in the market estimation of the marginal efficiency of different types of capital, calculated on the principles I have described above, will be too great to be offset by any practicable changes in the rate of interest.

Disappointingly, he did not investigate the role of innovations in generating a revival of animal spirits and raising the level of expectations for future profits. But the aphrodisiac effect of a wave of new investment opportunities based on a cluster of innovations is quite consistent with his general approach to expectations and investment behaviour. His recognition that innovative investment had almost as little to do with mathematical calculation as an expedition to the

South Pole does not invalidate this point. While pure animal spirits often sustain the early pioneers of a new technology (as we shall see when we come to discuss in more detail the rise of new industries), the swarming of imitators is usually promoted by the example of one or more exceptionally profitable ventures. Like those who joined in the 'Gold Rush' of the nineteenth century many of them are doomed to disappointment, but their 'swarming' behaviour is far from being entirely irrational, even though fashionable trends may play a part in influencing behaviour.

The effects of the volatility of attitudes and expectations (for whatever causes) led in Keynes' view to the exaggeration of both slumps and booms and to big variations over time in the rate of capital accumulation. However, he was more interested in the short run than the long run and did not really examine some of the longer term consequences of these fluctuations. The process of capital accumulation is affected not only by the vagaries of investment behaviour which Keynes described but also by a variety of inflexibilities affecting the growth of the capital stock, such as:

(i) inflexibilities in terms of delays in perceiving the need for new fixed capital and in the necessary time lag before the new fixed capital can be ordered, designed and constructed;
(ii) inflexibilities in terms of the longevity of capital;
(iii) inflexibilities in terms of the possibilities of factor substitution between capital and labour.

These inflexibilities will not only lead to disequilibria, but might well result in cyclical crises. Long-run capital accumulation is therefore at the core of many long-wave theories. The first of these inflexibilities can lead to cyclical over- and under-expansion of the capital-producing sectors, and is most easily identifiable with the traditional multiplier–accelerator models and most of the work of Forrester (1981) and the MIT System Dynamics Group. The second type of inflexibility is important in that it raises the possibility that large investments in one period, by becoming obsolete at the same time, will produce a flurry of replacement investment activity which may recur periodically (the so-called 'echo' effect). This was originally suggested by Kondratiev himself as a possible mechanism by which long waves are generated. Its effect might have been of some relevance to the post-war economic recovery, with its huge replacement and expansion of housing, public transport, motorways and other public works. It might also be possible that, in some countries, much of the fixed capital associated with this heavy post-war investment may need to be replaced over a rather short period in the future.

The third type of inflexibility is, in its extreme form, reflected in the assumption of 'fixed coefficients of production' as used, for example, in many so-called 'vintage' models. Most of these models assume that capital/labour substitution possibilities are confined to *ex ante* investment decisions (putty), whereas *ex post* the possibility of substitution disappears (clay) (i.e., any additional labour working with the existing capital stock will be purely un-productive, its marginal product being nil). In other words, one assumes that businessmen face a series of investment possibilities (the latest vintages) defined in terms of investment/labour ratios. Once the choice is made and the investment installed, changes in factor prices (e.g., wage increases) will not permit a change in the production factor content of the investment. The only option left is prematurely to scrap the investment. While this third inflex-ibility should not necessarily lead to any cyclical pattern (though such a pattern might exist (Clark 1980)), it indicates very clearly the crucial role of technical change and factor price 'expectations' (e.g., future wage settlements). Not surprisingly most of these vintage models arrive at the conclusion that the increasing labour share of output has led to a reduction in the life-time of the capital stock (e.g., see Vandoorne and Meeusen, 1978), and the ensuing de-cline in employment. The possibility that recent low levels of invest-ment may lead to capital shortage unemployment, once expansion of output gets underway, is discussed in Chapters 8 and 10.

As will have become obvious from this last discussion, technical change plays a crucial role in creating some of these inflexibilities, lags or discontinuities. In terms of short-run investment demand, new technology should normally increase expectations (i.e., would normally have a net employment-generating effect). However, changes in the structure of investment demand because of new tech-nologies might lead to an increased price for the new capital good or the associated 'software', which might offset some of this increase in expectations. Long-run capital accumulation on the other hand (as emphasized by Neisser, 1942) can have employment displacing as well as employment-generating effects. This will primarily depend on the nature of the investment (expansion or rationalization), relative factor prices and their expected trend.

As in the case of the capital stock, the labour market also suffers from a number of rigidities which prevent immediate adaptation to changes in industrial structures that result from factors such as technical change and changing demand patterns. There are clearly good social and cultural reasons for many labour inflexibilities. Job insecurity or geographical mobility do have a social cost. In terms of rapid technical change, changes in the demand for certain skills (which are in short or inelastic supply) will also create bottle-necks. If in addition, by the time reskilling or retraining has increased the supply of these scarce labour skills, technical change has again changed the demand pattern, one might well end up with a large number of disequilibria in the labour market. These disequilibria can be further exacerbated by wage differential inflexibilities, and by a more general 'stickiness' in wage rates, engendered in part by institutional factors in the wage bargaining process and by lags in the adaptation of bargaining attitudes to changed circumstances.

2.4 Equilibrium and structural change

The more or less 'traditional' explanations of economic development tend to assume that there exists some sort of general equilibrium growth path. We have listed a number of factors: time lags in the adaptation of the capital stock; undetermined expectations; interest inelasticity of investment demand; limited factor substitution, etc., which, combined with wage and price inflexibilities, might lead to a sustained departure from that equilibrium. We also insisted on the role of technical change in creating possible further departures from that equilibrium. Most of that debate (in particular the technical change bias) forms now a central part of economic growth textbooks. However, the way most of these questions are put (i.e., in terms of general equilibrium), make them of very little use in explaining major fluctuations as it is implicitly assumed that departures from the equilibrium path are short-lived frictional imperfections. A steady state growth path, a dynamic full employment general equilibrium, a constant rate of 'Harrod-neutral' technical change are all concepts, which — though useful from a theoretical point of view — at a certain stage seem more to obscure than to clarify the actual interactions between growth, technical change and employment.

This is where in our view Schumpeter's main contribution lies. In Schumpeter's framework it is disequilibrium, dynamic competition (in the sense of 'imperfect' competition) among entre-preneurs, primarily in terms of industrial innovation, which forms the basis of economic development. Thus, the emphasis is on the supply side, that is, autonomous investments rather than on 'demand induced accelerator investments or multiplier processes [demand push] as driving forces in economic development' (Giersch 1979, p. 630). In such a framework economic development will be viewed primarily as a process of reallocation of resources *between industries*.

That process leads automatically to *structural* changes and dis-equilibria, if only because of the uneven rate of technical change between industries (see Chapter 7). In other words, economic growth is not only accompanied by the rapid expansion of new industries, it also primarily depends on that expansion. It is not difficult to see how that expansion can lead to crises.

First of all, as emphasised by Kuznets (1930 and 1954), by Burns (1934), and by much of the early marketing literature and later the product life cycle trade theory, there is nothing steady about the expansion of products or industries. Rather it seems more plausible to postulate some sort of 'cyclical' expansion, with a fast growing, saturation and declining phase as the most obvious ingredients. Second, as pointed out by van der Zwan (1979, p. 23):

the lagging behind the trend of economic improvement by the industries out-side the cluster of modern and advancing ones leads to an increase in prices in this 'backward' sector. As a direct consequence the advancing sector is con-fronted with rising costs for its inputs, which leaves it with additional demands as far as productivity is concerned. An indirect consequence of the relative price increase in the 'backward' sector is the negative impact on purchasing power in the economy at large; this hampers the expansion of the advancing sector on the side of demand.

In other words, because of the productivity differentials between industries, the fast-growing sector of the economy will face demand problems and its expansion will be hampered and retarded at some stages of its growth. Third, to the extent that technical change is size biased, and that there might be a tendency in the fast-growing sectors of the economy to over-anticipate demand, there might also ultimately be a tendency to overproduction and excess capacity in the modern sectors of the economy. In Chapter 7 we take up the discussion of these productivity differentials between industries and to the question of 'saturation' of consumer demand for par-ticular commodities. As Pasinetti (1981) has argued, the expansion of consumer demand is itself not a smooth continuous process.

Within a Schumpeterian framework, structural 'adjustment' dis-equilibria are the logical outcome of economic development. The ways in which these disequilibria lead to 'long waves' and what might be the role of technical innovations in creating or amplifying these waves is discussed in the following chapters.

From what has been said it is evident that we believe there are good reasons for viewing technical change as one of the main factors involved in the various structural crises which have affected the growth of the world economy over the past two centuries, especially the level of employment. We do not, however, subscribe to a theory of single-factor causation, nor to a conception of regular repetition with fixed periodicity of an unchanging cyclical mode of development. On the contrary, whilst we would go along with Schumpeter in his insistence on the central role of technical change in the dynamics of capitalist development, we also share his view, expressed so clearly at the beginning of his analysis of *Business Cycles* (see Fels 1964, pp. 11–12):

> . . . the question of causation is the Fundamental Question. . . . Now if we do ask this question quite generally about all the fluctuations, crises, booms, depressions that have ever been observed, the only answer is that there is no single cause or prime mover which accounts for them. Nor is there ever any set of causes which account for all of them equally well. For each one is a historic individual and never like any other, either in the way it comes about or the picture it presents. To get at the causation of each we must analyse the facts of each and its individual background. Any answer in terms of a single cause is sure to be wrong.

In his insistence on the importance of the specific features of each perturbation to the system, Schumpeter differed from most Keynesian theorizing and from much other macro-economic crisis theory with its blithe disregard of actual technical developments and real changes in the structure of industry and services. Like Marx and Kuznets his approach was evolutionary and historical and he was constantly aware of the limitations and dangers of abstract generalizations and models. He saw research on 'business cycles' as having to rely on company histories, technical journals, studies of particular products and branches of industry, and repeatedly emphasized how misleading aggregative statistics could be, since they frequently concealed rather than revealed the underlying processes of change.

He justified his view that technical innovation was 'more like a series of explosions than a gentle though incessant transformation' on three grounds. First, he argued that innovations are 'not at any time distributed over the whole economic system at random, but tend to concentrate in certain sectors and their surroundings', and that consequently they are 'lop-sided, discontinuous, dis-harmonious by nature'. Secondly, he argued that the diffusion process was inherently a very uneven one because 'innovations do not remain isolated events, and are not evenly distributed in time . . . on the contrary, they tend to cluster, to come about in bunches, simply because first some and then most firms follow in the wake of successful innovation'. Thirdly, he maintained that these two characteristics of the innovative process implied that the disturbances it engendered could be enough to 'disrupt the existing system and enforce a distinct process of adaptation' (see Fels 1964, pp. 75–7).

Hardly anyone would deny the truth of Schumpeter's first proposition. It is confirmed by a great deal of empirical research on the uneven distribution of R and D, patents, inventions and innovations between the various branches of the economy. The differences between rates of growth of various branches of production are well-known and obvious, as is the fact that some industries decline whilst others grow rapidly. Moreover, it is also universally agreed that many of these structural changes are the result of technical innovation. The decline of the canals and horse transport and the rise of the railways is an obvious case, followed by the rise of the internal combustion engine and the decline of the railways. Changes in the pattern of energy production and distribution are another related and obvious case. No-one would deny that the social and economic changes arising from these major processes of technical innovation were sufficient to entail substantial problems of structural adaptation, especially for those countries which already had a large capital stock and pool of skilled labour devoted to the exploitation of the older systems of technology.

More comprehensive evidence is now available which shows that differences in growth rates of production and of productivity have been systematically related to R and D intensity and to patterns of technical change (Terleckyj 1974). The most R and D intensive industries are those with extraordinarily high rates of growth in the twentieth century and most of them did not exist at all before this century — electronics, aircraft, drugs, scientific instruments, synthetic materials (Freeman 1974). For long periods growth rates of 15 per cent per annum in many industrialized countries were commonplace for these branches of industry. Again, it is fairly obvious that these high growth rates were related to a much greater flow of technical innovation in new products and processes and the high rate of diffusion of these innovations through the world economy. There has been a very high concentration of total R and D in electronics and chemicals in almost all the OECD countries in the past thirty years.

At the other extreme are industries in a process of decline or showing rather low rates of growth. These are frequently characterized by low R and D intensities (sometimes zero) and a low rate of technical change. Where technical innovation is taking place, as for example, in the textile and printing industries, it is often due more to the impetus from suppliers of machinery and materials, than from the efforts of the industry itself. The existence of a statistical association between measures of technical change and the growth of an industry or product group does not, of course, necessarily mean that it is technical innovation which has caused the growth. The reverse could be true, or both could be ascribed to some other factor, such as the quality of entrepreneurship or market demand. Schumpeter stressed the importance of autonomous invention and entrepreneurship, but Schmookler (1966) put the main emphasis on market demand.

2.5 Schumpeter or Schmookler?

However, if market demand were the only problem, then technical innovation could be regarded as a secondary phenomenon and taken for granted, since it would simply respond to demand management. It would be part of the adjustment to changing patterns of demand, not a semi-autonomous engine of growth. Schmookler's theory has sometimes been interpreted in this way. Fluctuations in investment are followed by fluctuations in inventions. In Chapters 5 and 6 we shall present some evidence which suggests that a purely demand-led theory of invention and innovation does not correspond to the historical facts in the case of the two technologies which we discuss. Schumpeter's theory of an autonomous impetus on the supply side deriving from advances in science and invention and realised through imaginative entrepreneurship appears to fit the facts rather better. Once, however, a major innovation has been made, then a pattern of demand-led secondary inventions and innovations may set in over many decades giving apparent credibility to a 'Schmookler'-type of analysis.

Dosi (1982) has suggested an interesting parallel between technological paradigms and Kuhn's theory of scientific paradigms. A major new technology is comparable to a new paradigm in science; as the technology takes off its further articulation is the outcome of 'natural trajectory' possibilities (comparable to normal science in Kuhn's theory) heavily influenced by the market selection environment. Technical change is both an engine and a thermostat, but the thermostat function tends to predominate as a technology matures.

It may be maintained that to cite Schmookler's (1966) work as an example of a demand-led theory of invention and innovation is an illegitimate over-simplification. He is, himself, at pains in the early part of his book to emphasize that basic science may have an important independent influence and unlike Hessen (1931) he does not argue that basic science is also demand-led. On the contrary he uses the metaphor of the two blades of a pair of scissors to describe the process of technical change, one blade being scientific discovery and the other one the changes in the state of market demand. However in practice he concentrates almost entirely on the demand 'blade' and although he does devote some attention to the distinction between important patents and secondary patents, in the end he minimizes its significance.

Despite his early disclaimer, demand leadership of invention is the most powerful message appearing to emerge from Schmookler's work, as will be evident from his own conclusions:

> The possibility that the results reflect the effect of capital goods inventions on capital goods sales is grossly implausible. In the time series comparisons, trend turning points tend to occur in sales before they do in patents and long swing troughs in sales generally precede those in patents. Moreover, trends and long swings in investment in the industries examined are adequately explained on other grounds. . . . The fact that inventions are usually made because men want to solve economic problems or capitalise on economic opportunities is of overwhelming importance for economic theory. Hitherto, many economists have regarded invention − and technological change generally − as an *exogenous*, and some even thought, an *autonomous*, variable. . . . These views insofar as they were of a substantive nature rather than merely a methodological convenience . . . are no longer tenable . . . the production of inventions, and much other technological knowledge, whether routinised or not . . . is in most instances as much an economic activity as is the production of bread (1966, pp. 204–8).

Figure 2.2 gives a schematic representation of Schmookler's theory. Initially, a small increase in demand might be met by increased production from existing plant (Route 1) or with a time lag through expansion of capacity using existing technology (Route 2). More commonly however, given strong demand, new investment would generate an increase in inventive activities both within existing firms and outside them (Route 3). Expansion of investment will give rise to an increase in the rate of invention and the number of patents. On a very small scale in the nineteenth century, but much more generally in the twentieth century, efforts to improve the rate of invention and to appropriate the results monopolistically led to the growth of company-financed ('captive') R and D laboratories and other technical services (Route 4). The new inventions and improvements would be embodied in existing and in new production facilities and in new and improved products to meet the increased demand. Cyclical falls in demand would lead to corresponding reductions in inventive and other R and D activities.

Schmookler succeeded in demonstrating a high degree of synchronicity in the movements of his long time-series on sales of capital goods (or investment) and patented inventions in the same

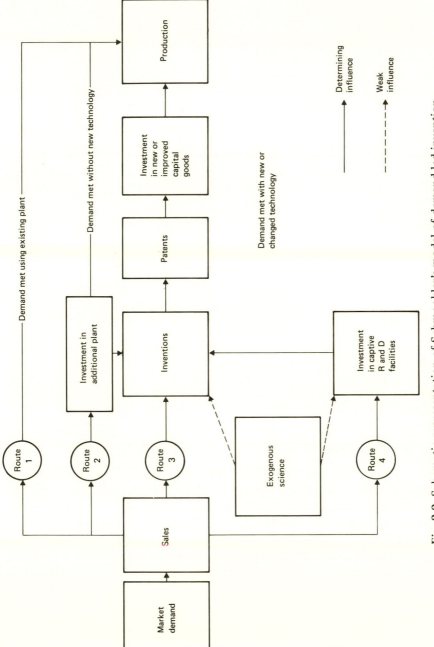

Fig. 2.2 Schematic representation of Schmookler's model of demand-led invention

field, especially in relation to long (Kuznets)* cycles. He was, however, less successful in demonstrating any consistent time lag between the two series, for as he frankly admitted, in some cases there were occasions when patents appeared to lead investment, as well as the (slightly more frequent) cases of investment leading patents. He based his argument therefore not on a general consistent lead of the sales and investment series, but on the argument that investment usually led the upswings from the troughs, and on the view that the movements of the investment series could be better explained by external events other than the course of invention.

In Chapter 5 we conclude that the evidence from the plastics industry, whether descriptive or statistical, does not support either a deterministic model of demand-led invention or of technology-led science. Neither would it sustain a theory of pure 'discovery push' or invention push, which ignored the reciprocal influence of the growth of demand, of fluctuations in economic activity and of competitive pressures. But it would be consistent with a theory such as Schumpeter's which postulated a closely interdependent but shifting relationship changing over time as an industry grew to maturity. Exogenous science and new technology tend to dominate in the early stages, whilst demand tends to take over as the industry becomes established. A 'matching' process of new technology and new markets, guided by imaginative entrepreneurs, is important throughout.

However, as Almarin Phillips (1971) has cogently pointed out there is not one Schumpeterian model but two. The first is that already developed by the young Schumpeter before World War II and expounded in his *Theory of Economic Development* (1912). The second is that advanced in his later book *Capitalism, Socialism and Democracy* (1943). Figures 2.3 and 2.4 are a schematic representation of these two models which we shall designate as Schumpeter I and Schumpeter II. They are based essentially on the diagrams used by Phillips (1971) with minor modifications. The pattern postulated by the Schumpeter I model and illustrated in Figure 2.3 may be summarized as follows:

(i) There is a (discontinuous) flow of basic inventions related in an unspecified way to new developments in science. These are largely exogenous to existing firms and market structures, and hence to any measurable type of 'market demand', although they may certainly

*Historical studies of the USA economy have often been based on the long cycles identified by Kuznets, which he associated with long-term trends in building and construction.

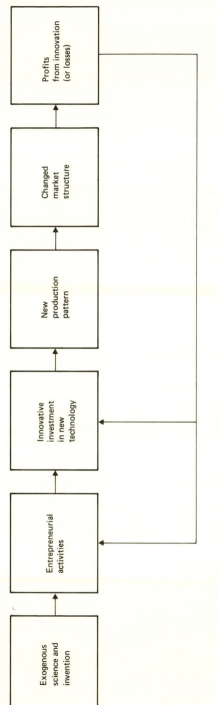

Fig. 2.3 Schematic representation of Schumpeter's model of entrepreneurial innovation (mark I).

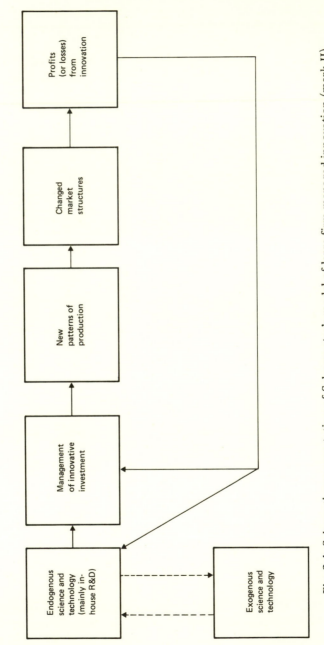

Fig. 2.4 Schematic representation of Schumpeter's model of large-firm managed innovation (mark II).

be influenced by the belief in a potential demand or concept of unmet need, or shortages of existing products.

(ii) A group of entrepreneurs (who in Schumpeter's view are responsible for the main dynamic thrust in capitalist economies) realize the future potential of these inventions and are prepared to take the risk of developing and innovating. This hazardous activity would not be undertaken by the average capitalist but only by exceptional entrepreneurs.

(iii) Once a radical innovation had been made it would disequilibrate existing market structures and reward the successful innovator with exceptional growth and temporary monopoly profits. However, this monopoly will be later whittled away by the entry of swarming secondary innovators giving rise to the cyclical phenomena already described.

The main differences between Schumpeter II and Schumpeter I are in the incorporation of *endogenous* scientific and technical activities conducted by large firms. Schumpeter did not have industrial R and D statistics at his disposal as these have only been systematically collected since World War II. However, he was undoubtedly aware of the rapid growth of these activities between the wars and the extent to which innovative activity was generally becoming institutionalized within large corporations. Indeed, he was so impressed by these tendencies that he foresaw the innovative entrepreneur as ultimately being completely superseded by a 'bureaucratized' type of innovation. He had actually foreshadowed this change in the emphasis of his theory in an article as early as 1928. In *Capitalism, Socialism and Democracy* (1943) he went even further and saw this tendency as the main force which would ultimately lead to the disappearance of capitalism itself.

In Schumpeter II (Figure 2.4) therefore there is a strong positive feedback loop from successful innovation to increased R and D activities setting up a 'virtuous' self-reinforcing circle leading to renewed impulses to increased market concentration. Schumpeter now sees inventive activities as increasingly under the control of large firms and reinforcing their competitive position. The 'coupling' between science, technology, innovative investment and the market, once loose and subject to long time delays, is now much more intimate and continuous.

Whilst some economists have seen Schumpeter's models as essentially two conflicting views of the world, we prefer to regard them as complementary, and reflecting not so much a different analysis as a different world which was being analysed. Model I was based on nineteenth-century history as seen in the early years of the twentieth century. Model II was based on what had been happening in World War I and the inter-war period.

Although our evidence appears to support the Schumpeterian model (and those theorists like Phillips, Rosenberg and Nelson who have followed largely in his tradition), there is one important line of research and argument which might appear to refute Schumpeter II whilst continuing to vindicate Schumpeter I. This is the argument that large firms tend to rest on their monopolistic laurels, to stagnate and to become less innovative, thus providing continuous new opportunities for the small innovative firm to undermine the large monopolistic firms despite the scale of their R and D and other technical resources.

Especially in the birth of new industries, small firms may be exceptionally important. In the 1950s it was often argued that the process of industrial concentration had slowed down or ceased, and also that the largest firms tended to be relatively less R and D-intensive than small or medium-sized ones. Scherer (1965) and Hamberg (1966) in particular argued from the statistical evidence then available that there was no association between firm size and relative scale of research and inventive activities, or if there was such an association then it was one in which R and D efforts tended to diminish in the largest firms.

However, this argument has been overstated. The evidence of growing industrial concentration in the 1960s and 1970s is now seldom disputed, but the Schumpeterian (Mark II) explanation is often ignored or contradicted, usually on the basis of Scherer's work. This view was based largely on the statistics of the 1950s which were unsatisfactory and provided no time-trend. In a recent paper on the subject Soete (1979a) made use of statistics of the National Science Foundation for the 1970s and other US sources to show that Schumpeter Mark II was largely valid. In most branches of US industry, including chemicals, it is the largest firms which are the most R and D intensive. Another study in the UK for the Bolton Committee (Freeman 1971) showed that large firms accounted for a disproportionately large share of post-war British innovations, and that in many industries scarcely any significant innovations came from small firms (see also Townsend et al., 1981).

A general long-term tendency towards concentration of innovations in larger firms is quite consistent with the possibility that long-term cyclical upswings are associated with a resurgence of 'Mark I' small-firm innovations and, as we shall see in Chapter 6, the pattern in electronics and scientific instruments does provide evidence of this. In Chapter 7 we present further evidence both on the long-term tendency towards concentration of production and sales, and on the resurgence of fast-growing small innovation firms in such new technologies as information processing and small computers.

Obviously these Schumpeterian models represent a rather different view of the determinants of technical change and economic development than that propounded by Schmookler. However, it is important to note that in terms of Schmookler's statistical analysis the results might sometimes look the same. Since little distinction is made between the original radical inventions and innovations and all subsequent inventions in Schmookler's method, in terms of simple patent counts, there would always be far more inventions and patents by the 'swarming' secondary innovators than by the original radical inventors and innovators. Moreover, these might often tend to follow rather than precede the upsurge or decline of investment and production. Thus the Schumpeterian model is not necessarily inconsistent with the Schmookler statistics. Its validity can be tested only by the descriptive narrative type of approach which we have attempted in our discussion and detailed case studies in Chapters 5 and 6, as well as by a statistical analysis which attempts to separate basic inventions and innovations from the rest. In Chapter 3 we develop this argument in relation to inventions and innovations generally. Engines must be distinguished from thermostats.

3 MENSCH'S THEORY OF 'BUNCHING' OF BASIC INNOVATIONS

As we have seen, in his review of Schumpeter's *Business Cycles* Kuznets (1940) maintained that the argument would stand or fall on the possibility of demonstrating the existence of discontinuities in the introduction of major innovations with a long time span. Minor technical improvements were being made all the time in many sectors of the economy, but these would be incapable of generating the type of fluctuations suggested by Schumpeter.

In view of these comments, particular interest attaches to the attempt by Mensch (1975) to demonstrate precisely such discontinuities in relation to what he describes as *Basis-innovationen* (basic or radical innovations). In his view they have tended to cluster in a few decades over the past two centuries (*viz.* in the Depression periods of the 1830s, 1880s and 1930s). He accepts that depressions need not and should not be the only way to get a flow of basic innovations (p. 168) and indeed he argues strongly for an active industrial policy to do it without. But he maintains that historically as a matter of actual fact, basic innovations *have* clustered in and around depressions. He also predicts a new cluster of basic innovations in the 1980s.

There are two main aspects of this Mensch theory of innovations. First, he argues that a 'technological stalemate' period tends to follow the peak of a long wave boom. In this period of stagflation or recession *Verbesserungs-innovationen* (improvement innovations) and *Schein-innovationen* ('pseudo-innovations' or product differentiation) tend to proliferate, concentrating on modifications to the existing industrial system. This part of his argument is more plausible, although it merits more empirical investigation than it has so far received, and is open to criticism on the grounds of neglecting basic process innovations, for which there would still be a strong incentive in periods of recession. The second part of his theory relates to *Basis-innovationen* (basic or radical innovations). He believes that these are crowded out and tend to diminish in number during the stagflation periods but that major depressions remove the barriers to basic innovation. They are brought forward as the result of an 'accelerator mechanism' which shortens the lead time between invention and innovation during deep depressions. Hence they tend to bunch together in the decades of deep troughs in the economic system and provide the stimulus and opportunity for the subsequent recovery and boom.

In our view the significance of 'swarming' or of 'bunching' relates more to the diffusion process and to combinations of sets of basic innovations with many improvement innovations rather than the initial appearance of the individual basic innovations themselves. We shall develop this approach in Chapters 4, 5 and 6. In this chapter we concentrate on a critique of the Mensch theory. The reason for devoting considerable attention to this critique is that the ideas have been widely discussed and quoted and have been an influential factor in the wider debate on innovation and economic growth. In particular the main argument has been taken up by Jay Forrester and his colleagues in discussing the long-term growth of the US economy. Graham and Senge (1980) explicitly incorporate Mensch's theory into their own analysis. However, they put their main emphasis on basic innovations in the 'early upswing' period from depressions (p. 13). Mandel (1981) also accepts the validity of Mensch's findings. Finally, in our view a critique of Mensch's pioneering work aids the process of clarification of the main issues.

3.1 The bunching of innovations in the twentieth century

Unlike some other critics we do not disagree with Mensch in his attempt to distinguish between 'basic innovations' and 'minor improvements'. On the contrary we think that he is right to make this distinction and has not gone far enough in this direction. However difficult it may be in practice to classify the different types of invention and innovation, or to measure them and to assess their trends over time, there really *is* an important difference between the innovation of nylon or the electronic computer and a technical improvement to the horseshoe or the tin-opener, even if there were as many patents for the latter two as the first two in any particular year (See Appendix A, on p. 201 for various definitions of invention, innovation and diffusion.)

In his book Mensch relies heavily on a list of forty-one basic twentieth century innovations derived from the classic work of Jewkes *et al.* (1958) *The Sources of Invention.*

Using the case studies at the end of this book Mensch attempts to date these innovations according to the year in which production began on a factory scale or a new market was established. From this analysis he concludes that only one of them was introduced before 1920, seven during the 1920s, twenty during the 1930s, eight in the

1940s, and five in the 1950s. This appears to fit well with the notion of depression-induced bunching. He uses some other sources which are less supportive of his main conclusions than the list of Jewkes *et al.* but criticisms similar to those below apply to them too, especially the Schmookler lists (1966).

It is important to recognize that Jewkes *et al.* never claimed that their selection of sixty-one inventions in the first edition of their book (or seventy-one in the second edition) was in any sense complete or a statistically reliable sample of inventions. Moreover there are several obvious sources of bias for the purposes in which Mensch is interested. He relies on the first edition and since this was compiled in the mid-1950s, it is obvious that it could not do justice to the inventions and innovations of the 1950s or the 1960s. In their second edition Jewkes *et al.* point out that they omitted the computer from the first edition because they did not realise how important it would be, and they had no idea how important the transistor would be either. There is a similar problem at the beginning of the century since Jewkes *et al.* refer to inventions, whereas Mensch is interested in innovations. Thus the aeroplane invented at the turn of the century does not appear on either list whereas helicopters do. Using the list of Jewkes *et al.* has probably resulted in considerable under-estimation of the innovations of the early part of this century as well as those of the post-war period.

Mensch's purpose was not himself to construct a list of twentieth-century innovations but rather to use three different sources to show that all of them quite independently demonstrated a bunching effect in the 1930s. But Schmookler's list is even more deficient for his purpose than that of Jewkes *et al.* It related to four industries only and three of them were already 'old' industries for the twentieth century — agriculture, railways and paper. Moreover, his study stopped specifically in the mid-1950s. If, therefore, two of his sources are not suitable for the purpose which Mensch intends, the statistical argument is greatly weakened. The probability that there are some bunching effects in relation to the flow of basic innovations over time certainly should not be excluded and we shall be presenting shortly the results of some other empirical work on the bunching of both inventions and innovations, but Mensch's empirical evidence cannot be regarded as adequate to support his conclusions.

There are a number of further subsidiary points which raise additional doubts about the validity of the use of the list of Jewkes *et al.* for his main purpose.

Some twenty-two inventions from the list in the first edition of

Jewkes *et al.* are not included. Assuming that Mensch is relying only on the case studies, this number falls to eleven. The chief criterion for exclusion is apparently that the innovations concerned have an estimated 'lead time' between invention and innovation of ten years or less (Mensch 1981) but it is not at all clear why an innovation with a relatively short lead time should be considered less important for the analysis. Most of those innovations omitted would fall outside the 1930s and add to the numbers before 1920 and after 1940 (see Table 3.1 part (b)). The ten innovations added in the second edition would all add to the numbers after 1940 (Table 3.1 part (c)) thus weakening the 'bunching' of the 1930s identified by Mensch.

A second point with respect to interpretation is that there is a high degree of ambiguity surrounding any estimate of invention and innovation dates, and differences of opinion are likely even when a single source of information such as Jewkes *et al.* is used. Some of those classified to the 1930s in Mensch's list could equally plausibly be allocated to other decades — for example, the diesel electric locomotive (1913) or ballpoint pen (1945).

The three of us have independently estimated the dates of invention and innovation for all sixty cases covered in the case histories sections of the two editions of Jewkes *et al.* and have found that in a number of them we disagreed between ourselves and with Mensch. We have, however, arrived at a joint 'consensus' estimate in each case, and compared our results with those of Mensch. We do not maintain that our estimates are better or more realistic, but merely that they are equally plausible alternatives. These are shown in Table 3.1.

In any attempts to make a more satisfactory analysis there is another problem which must be taken into account — the problem of component innovations, system innovations, and 'families' of innovations. On Mensch's list there are two cotton-pickers and two forms of automatic transmissions, whereas Jewkes himself counts each of these as only one. There is a problem of consistency in counting the cotton-picker as two but radio or the transistor as one each. Chapter 6 lists more than thirty innovations in solid state technology in the 1950s and 1960s, and a number of these would rate as basic innovations as well as the transistor itself.

Finally, there is a problem of weighting *within* the category of basic innovations. The list derived from Jewkes includes the zip fastener but not nuclear reactors, includes insulin but not the pill or tranquillizers, includes the ballpoint pen but not the electronic computer, neoprene but not the Haber–Bosch process. Again,

Table 3.1   List of innovations as reported in Jewkes, Sawers and Stillerman: Estimated invention and innovation years; where these differ from Mensch, the *latter's* estimates are placed in brackets

New concept	Invention	Innovation
(a) *Cases considered by Mensch (1975)*		
Automatic drive	1904	1939
Hydraulic clutch	1924(1904)	1946(1937)
Ballpoint pen	1888	1946(1938)
Catalytic cracking of petroleum	1915	1935
Watertight cellophane	1912(1900)	1926
Cinerama	1937	1953
Continuous steelcasting	1927	1952(1948)
Continuous hotstrip	1920(1892)	1923
Cotton picker (Campbell)	1920	1942
Cotton picker (Rust)	1924	1941
Crease-resisting fabric	1926(1906)	1932
Diesel locomotive	1895	1913(1934)
Fluorescent lighting	1901(1852)	1938(1934)
Helicopter	1904	1936
Insulin	1920(1889)	1927(1922)
Jet engine	1928	1941
Kodachrome	1921(1910)	1935
Magnetic taperecording	1898	1937
Plexiglass	1912(1877)	1935
Neoprene	1906	1932
Nylon, perlon	1927	1938
Penicillin	1928(1922)	1943(1941)
Polyethylene	1933	1937(1953)
Power steering	1925(1900)	1930
Radar	1925(1887)	1934
Radio	1900(1887)	1918(1922)
Rockets	1923-9(1903)	1942-4(1935)
Silicones	1910(1904)	1943(1946)
Streptomycin	1942-3(1921)	1944
Sulzer loom	1928	1945
Synthetic detergents	1886	1928
Gyrocompass	1900(1827)	1909
Synthetic light polarizer	1928(1857)	1932
Television	1923(1907)	1936
'Terylene' polyester fibre	1941	1955
No-knock gasoline	1912	1935
Titanium	1937(1885)	1944(1937)
Transistor	1948(1940)	1950
Tungsten carbide	1900	1926
Xerography	1937(1934)	1950
Zipper	1891	1923

New concept	Invention	Innovation
(b) *Cases analysed by Jewkes et al. but excluded in Mensch (1975)*†*		
Bakelite	1904	1910
Cyclotron	1929	1937
DDT	1874(1939)	1942
Electric precipitation	1884	1909
Freon refrigerants	1930	1931
Hardening of liquid fats	1900	1909
Long-playing record	1945	1948
Safety razor	1895	1904
Self-winding wristwatch	1922	1928
Shell moulding	1941	1948
Stainless steels	1904	1912
(c) *Additional cases listed and analysed in Jewkes, Sawers and Stillerman,.* *2nd Edition* (estimates by current authors)		
Hovercraft	1955	1968
Chlordane, aldrin, dieldrin	1944	1946/47
Float glass	1902	1957
Moulton bicycle	1959	1963
Oxygen steel-making	1949	1952
Electronic digital computers	1939	1943
Photo typesetting	1936	1954
The preventionof rhesus haemolytic disease	1961	1967
Semi-synthetic penicillins	1957	1959
Wankel rotary piston engine	1957	1967

*The dates correspond to those cited in Mensch (1971) or for Bakelite and the long-playing record, in Mensch (1981). For DDT, our own dating of the invention differs substantially from Mensch's; the latter estimate is in brackets.

†In addition to these there were nine other inventions mentioned in the main text by Jewkes *et al.* but not included in their case studies at the end of the book.

recognizing that Mensch was not primarily concerned to make a complete list of basic innovations, the deficiencies of the lists he did use must cast some doubt on the conclusions.

Kleinknecht (1981) attempted to rescue the Mensch theory by the use of a new and apparently more recent and comprehensive list of innovations. Unfortunately, however, the book (Mahdavi 1972) containing this list is a compilation based on a series of earlier studies of innovation completed at various dates in the 1950s and 1960s. This is the only possible explanation of the fact that the last major drug innovation listed by Mahdavi (and Kleinknecht) is in 1948, when most other sources agree that there were many important drug innovations in the 1950s and 1960s (chlorpromazine, tetracyclines, the pill, paracetamol, librium, valium etc.), and indeed hardly any innovations are listed in any branch for the 1960s except in scientific instruments, where it is obvious that Mahdavi had access to a more recent study.

Thus the use of the Mahdavi list is actually open to exactly the same basic criticism as the use of the Jewkes and Schmookler lists — serious under-estimation of the innovations of the 1950s and 1960s, except in one category. Kleinknecht divides the list of 120 innovations into product innovations, improvement and process innovations and scientific instrument innovations. He bases his analysis on this classification and claims that the results of his testing show a very strong confirmation of the Mensch hypothesis for product innovations in depressions but rather weak support for the idea of bunching of process innovations in prosperity periods. Whilst this attempt to refine the original Mensch approach is welcome, there are great difficulties in categorizing and separating product from process innovations and there must be reservations about his separation of scientific instrument innovations from the rest. Mahdavi's data show a strong bunching of instrument innovations in the 1950s and 1960s. However, it could be argued that scientific instruments are just as much an industry as any other even though they do have special links with the R and D network. If instead of being separately classified they are added to either of the other two groups separately distinguished by Kleinknecht — whether the product category or the process category of innovations — then they greatly modify his results. In the first case they upset the 1930s bunching of product innovations because they cluster in the 1950s and in the second case they greatly strengthen the purely statistical evidence of a bunching of process innovations in the prosperity periods (1950s and 1960s), which he claims as otherwise only weakly supported by the Mahdavi data. The second result is particularly interesting in considering the strong evidence of the progress of automation in many industries in the 1950s, 1960s and 1970s, based on the introduction of novel process instrumentation associated with computers.

Van Duijn (1981), like Graham and Senge (1980), puts the emphasis on the bunching of innovations in the early recovery phase from a deep depression and accepts that the depressions themselves might actually have at least a temporary retarding effect on basic innovations. He also suggests that basic process innovations may tend to cluster later in the cycle and produces some evidence to support this view. Together with other refinements this seems a more acceptable version of the original Mensch theory, although it too remains to be more adequately tested with a larger sample of innovations.

In the Science Policy Research Unit we are attempting to develop a more systematic data bank on innovations, to enable a more thorough analysis of long-term trends. With the aid of a grant from the Joint Committee of the Science and Engineering Research Council and the Social Science Research Council, Townsend *et al.* (1981) have assembled information on more than 2000 innovations in about 60 per cent of the manufacturing industries of the UK. With their help and that of many technical experts in each of the relevant industrial branches we have distinguished about 200 radical innovations between 1920 and 1980 and dated them according to the first date of commercial application in the UK. The results are shown in Figure 3.1. It must be emphasized that they are provisional, that they cover only 60 per cent of industry and that the work is still in process.

As the graph shows, there was indeed a peak of innovation activity in the recovery period from the great depression in the late 1930s, but there was also a substantial continuing flow of radical innovations during the 1950s and 1960s, with a slight tendency for process innovations to increase more than product innovations. Finally our results do permit some confirmation of Mensch's theory of 'technological stalemate' — a decline of basic innovations, when the peak of the long wave expansion has passed. There is a marked fall in basic innovations both in the 1920s and the 1970s. However, the reservation must be repeated that any analysis of recent radical innovations probably underestimates them, because it may not be possible to assess their importance until later.

3.2 Depression as an accelerator mechanism for innovation?

We now turn to the second main point in the Mensch theory — the notion of an accelerator mechanism operating to reduce innovation lead times during periods of depression.

Figure 3.2, reproduced from Mensch (1975), shows the *invention* year plotted against the lead time from invention to innovation for the list he distilled from Jewkes *et al.* All these cases are estimated by Mensch to have become *innovations* between 1909 and 1955 inclusive. From the way the diagram is drawn, all the points representing innovations between 1909 and 1955 *must* lie between the dashed lines we have added to the diagram: the fact that two points lie outside these limits is apparently due to an arithmetic error in calculating the lead times for two types of cotton-picker (see Mensch 1975, Table 4.4). This diagram is then used by Mensch not only to support the idea of a 'bunching' of innovations

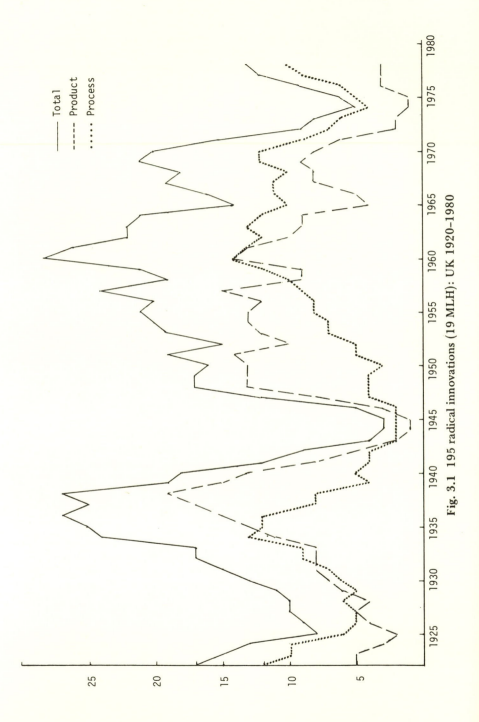

Fig. 3.1 195 radical innovations (19 MLH): UK 1920–1980

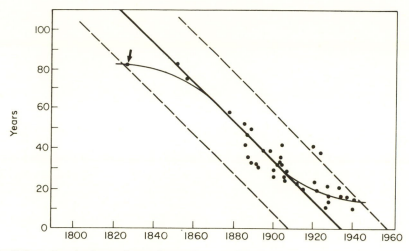

Fig. 3.2   Lead times from invention to innovation — Mensch's estimates. Source: Mensch (1975).

in the depression of the 1930s, but also to illustrate the effect of the 'accelerator' mechanism over that same period which drastically reduced innovation lead times.* In Figure 3.3 we have constructed a similar diagram on the basis of our estimates of the invention/innovation dates of the inventions discussed by Jewkes *et al.* This differs from Figure 3.2 in that:

(a) in some cases our estimates diverge from those of Mensch — these differences are listed in Table 3.1 part (a);
(b) we have included those innovations excluded in Mensch (1975) but included in Mensch (1971) Table 3.1 part (b);
(c) we have also included additional case histories covered in the second edition of Jewkes *et al.* (Table 3.1 part (c)).

*To assess the validity of these hypotheses, it is important to understand how the diagram is constructed. For any *invention*, the corresponding *innovation* will, by definition, occur T years later, where T is the lead time represented on the vertical axis. Consider, for example, the invention of the gyrocompass — the point arrowed in Figure 3.2, made according to Mensch in 1827 with an eighty-two-year lead time. The innovation year $(1827 + 82 = 1909)$ is the point where a line drawn in a south-easterly direction from the arrowed *invention* point on the diagram crosses the horizontal axis. Hence by definition, no points can lie outside the dashed lines drawn in Figure 3.2.

More formally, for a particular innovation, let t_1 denote the year of invention, t_2 the year of innovation and $T = (t_2 - t_1)$ the lead time. The co-ordinates of the point on the diagram representing this innovation are $(t_1, t_2 - t_1)$. The equation of the 'south-easterly' line (i.e., the line of slope -1) is $y = -x + c$, where the constant c can be found by substituting $x = t_1$, $y = t_2 - t_1$, since the line must go through this point. This gives $c = t_2$, so the line is $y = -x + t_2$. It hence crosses the horizontal axis ($y = 0$) at $x = t_2$, the innovation year.

With regard to the 'bunching' hypothesis, it is clear that a disproportionately large number of innovations fall in the 1930s (i.e., lie between lines with negative 45 degree slopes drawn from 1930 and 1940). However, we have already pointed out the unsuitability of the Jewkes *et al.* list for the purpose of testing Mensch's bunching hypothesis. In addition, we shall now argue that there is little evidence that those basic innovations that *did* fall within the period of depression were in fact induced by adverse economic circumstances.

The 'acceleration' hypothesis (i.e., the notion that lead times are reduced in periods of depression) would be reflected in these diagrams by a preponderance of points to the left of the solid diagonal lines in the upper parts of the diagrams, and a similar clustering to the right of this line in the lower parts. This would mean, for example, that in the earliest part of the twentieth century a disproportionate number of innovations had a comparatively long lead time, with an overall reduction in this lead time as the century wore on. It is clear that this hypothesis is very poorly supported by Figure 3.3; in Figure 3.2 its support depends to a great extent on the single point arrowed, representing the gyrocompass which, we believe, has been allocated an unrealistically early invention date by Mensch. Thus we regard the data of Jewkes *et al.* as providing no evidence of the 'acceleration' hypothesis, and, while superficially supporting the 'bunching' hypothesis, as unsuitable evidence for the conclusions that Mensch draws.

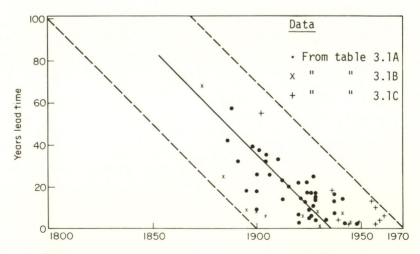

Fig. 3.3 Lead times from invention to innovation — authors' estimates

But even if the data were more reliable, simply to domonstrate that a 'bunch' of basic innovations precedes or follows an upswing or a downswing in a long wave or any other fluctuations in the economy does not in itself elucidate any causal mechanism which may be at work, even if (as Mensch maintains) the statistical probability of some association is extremely strong. To establish that depressions induce basic innovations or that booms may lead to a 'crowding out' of basic innovations or *vice versa*, or to demonstrate that a bunch of basic innovations could be one of the factors stimulating a long wave upswing, or to validate any other hypothesis about the reciprocal relationship between economic fluctuations and innovations, it would not only be necessary to collect good statistics and use them well, but also to show by the evidence of case studies that business (or government) behaviour was indeed influenced in the manner postulated by the theory.

Everyone accepts (including Mensch) that the gestation period for technical and commercial development of a basic invention varies a great deal depending upon the complexity of the problems, the resources committed to their solution, the efficiency of the R and D team, the quality of entrepreneurship, exogenous scientific and technical advances, competitor behaviour, the state of the potential market, legislation, the patent position, management decisions and so forth. Everyone agrees again that for basic innovations the gestation period may often be long — as much as 10–100 years. It is difficult to measure exactly because of the untidy and uncertain nature of innovation in general — projects may change their objectives as they go along as a result of unexpected discoveries; projects dropped by one individual or firm may be taken up by another; an innovation which has apparently failed in one field of application may be successfully launched in another and so forth.

To go through all the case histories of basic innovations of the past sixty years and check them for the operation of a 'depression-induced accelerator' mechanism would be a very time-consuming exercise. However, a preliminary test can be made from the case studies of Jewkes *et al.* from which Mensch drew up his list. None of the case studies mention that the depression of the 1930s influenced the decision making of innovators in a positive way, whether by the initiation of a new development project, by the acceleration of an existing project, by the earlier-than-expected launch of an innovation or by taking a completed or nearly-completed project 'off the shelf'. On the other hand no fewer than ten of the case studies (penicillin, hot-strip rolling, jet engine, radar, rockets, shell-moulding, silicones, titanium, tungsten carbide and computers) suggest positive effects on decision taking associated with strong demand in the recovery and prosperity phases, mainly rearmament and war but also the vehicle boom of the 1920s.

For example Jewkes *et al.* (1969) wrote that 'The outbreak of war stimulated the development of silicones to the production phase' (p. 298). Similar factors obviously applied in the case of radar and rockets (both dated by Mensch in the 1930s) and also to many synthetic materials, especially rubbers, and to penicillin. In the case of the influence of peace time demand the study of continuous hot strip rolling (1923) is interesting and Jewkes says that 'Hook was convinced that Tyne's ideas were sound and that the time was ripe to introduce them in view of the ever-increasing demands from the automobile industry for steel sheets' (p. 242). Two of the case studies mention the negative effects of depression on decision making (cotton-picker and power steering). Of the case of power steering Jewkes *et al.* say that 'the depression caused General Motors to abandon in 1933 their plans to introduce power steering' (ibid., p. 282). General Motors were not the first innovator in this field but as Mensch is counting this in his 1930s 'cluster', it would have to be explained in terms of the vehicle boom of the 1920s rather than the Depression of the 1930s.

There must be serious reservations, anyway, on the extent to which the acceleration of the gestation process of development is effective. Certainly there is some trade-off between cost and time, and military-crash projects have sometimes concentrated resources of high quality more quickly than would otherwise have been the case. However, clearly there are other cases where despite the best efforts of R and D teams specific technical problems could not be resolved. The 'cure' (or cures) for cancer and other malignant diseases are obvious cases in point. In these cases the limiting factor is not the availability of money or R and D resources, but the limitations of fundamental scientific knowledge. Basic research is an even more uncertain and unpredictable process than experimental development work and sometimes it is very hard to hurry it up. On the other hand, once a fundamental breakthrough is achieved it may open the flood-gates to a very large number of new technical developments. As we shall argue in Chapters 5 and 6, this seems to have been the case with macro-molecular chemistry in the 1920s through the work of Staudinger and Carothers, and with solid state physics in the 1940s and 1950s. It may also be the case with biotechnology at the present time. Consequently, we would disagree with Mensch totally in his rejection of the link between basic science and technology as one of the important mechanisms leading to the bunching of innovations.

This preliminary screening of the Jewkes *et al.* case studies is not a conclusive test or refutation of the hypothesis. The studies are highly condensed and more concerned with the personal experiences of the inventor and the technical problems of the innovation than with the influence of general economic conditions. However, if depression is as strong an accelerator mechanism on innovation as Mensch suggests, then at least some pale reflection of this mechanism would have been present in these summary case studies instead of the opposite. The fact that ten of the studies (eight of which Mensch includes in his list) mention in one way or another the positive effects of strong demand during rearmament, wartime and/or the prosperity phase suggests at least a *prima facie* case for the alternative hypothesis that this was a more important inducement mechanism than depression.

Chapter 2 has already made clear that we do not favour a simple demand-led theory of invention and innovation and we are not suggesting that only 'carrots' are effective in relation to basic innovations and that 'sticks' play no part. Actual or expected shortages have been demonstrably important inducement mechanisms for basic innovations in materials and energy and these may be construed as both 'stick' and 'carrot'. It can also be argued plausibly — as for example by Downie (1958) in a totally different context — that the 'stick' of failure or prospective failure in existing product lines should induce firms to attempt innovations. Since this stick will beat more people during a deep depression than at any other time, it could conceivably operate in the way which Mensch suggests. However, as against this the risks of radical innovative behaviour must appear even greater than usual during a deep slump.

Whilst accepting the point about the high risks of innovation during depression, Kleinknecht (1981) argues that the risks of *not* innovating may be even greater since some firms will be confronted with a situation where they *cannot* continue with their old product range. *Rien ne va plus*. They must either innovate or perish.

3.3 Long-term fluctuations in patents, inventions and R and D

The motives for innovation are many and varied and so are the opportunities. The most radical innovations depend in any case on rather unusual decisions. We know for sure that *some* basic innovations do occur at all stages of the medium- and long-term business cycles. Whether they will tend to occur rather more frequently during depressions cannot be determined by purely theoretical arguments about the balance of risks and the empirical evidence, although still inadequate, does not convincingly support the Mensch–Kleinknecht hypothesis.

There is however one further piece of relevant empirical evidence which does have a bearing on the controversy. Although we do not have comprehensive statistics of innovations, we do have more satisfactory statistics of inventions (patent statistics) and of R and D activities in industry. These are only indirectly related to innovation since the time lags between invention and innovation are variable and often long, and whilst the D (or development) part of R and D is directly concerned with innovation, the incidence of *basic* innovations (or even of minor innovations and technical improvements) cannot be deduced from fluctuations in development expenditures. However, variations in R and D activities and inventive activities *may* provide clues about the way in which firms respond to depressed business conditions. If there were evidence that firms responded to depression by stepping up their R and D activities, and increasing their applications for patent protection, then this would provide rather strong support for the Kleinknecht view of firm behaviour, even though these activities could not be directly related to the 'output' of basic innovations.

If, on the other hand, the empirical evidence suggests that firms respond to depression by cutting back on their research, inventive and technical development activities generally, then this must cast further doubt on the hypothesis of depression-induced acceleration of basic innovation, even though no direct and immediate relationships can be demonstrated and both basic inventions and basic innovations might still fluctuate independently and increase during depressions. What then is the evidence on the long-term trends in R and D activity and patenting?

During the post-war period it was often assumed that scientific and inventive activities were less vulnerable to the downswings of short- and medium-term business cycles than investment outlays. Generally speaking R and D expenditures for the OECD countries showed a strong and steady upward trend in the 1950s and 1960s, apparently little affected by short-term fluctuations. Whilst in the late 1960s there was a slowing-down of the previously very high rate of growth of these indicators, they still appeared to be little affected by business cyclical patterns. This was usually explained in terms of a recognition by firms of the essentially long-term nature of R and D and the need for stability in this sort of programmed activity. Whereas outlays for new machinery, vehicles and plant could be and were postponed or brought forward in response to changing business expectations, it was believed to be less sensible and less necessary to expand or contract R and D in response to the same pressures. Consequently, whereas investment behaviour occupies a central place in almost all theories of the business cycle, and indeed is the principal determinant of the cycle itself in many such theories, R and D behaviour is usually ignored. This must be qualified by the observation that R and D statistics only recently became generally available and that even now they are not always annual. The relative stability of R and D outlays during the 'recessions' of the 1950s and 1960s, moreover, does not mean that they are unaffected by deeper disturbances. Indeed, although we do not have comparable annual series for the pre-war period, there is very strong evidence that company outlays for R and D were significantly reduced during the depth of the depression from 1931 to 1934 in the leading industrial countries. Terleckyj (1963) reports a fall of more than 10 per cent in R and D in industry in the USA over this period. Furthermore, there is now additional evidence that in the more serious recent recessions of the 1970s and early 1980s R and D behaviour has been more affected than in the minor fluctuations of the previous twenty years. How about patents?

Figure 3.4 shows the total number of patents applied for and granted in the USA for the period from 1840 (year of the Patent Law) to 1979. The diagram suggests that variations in the number of patent *applications* have tended to coincide with overall long-term economic fluctuations. With an average of a four-year time lag (up to the 1960s, the average time needed for the Patent Office to examine and issue a patent was about four years) the evolution of the number of patents granted shows the same kind of behaviour. The steep fall in the early 1930s in patent applications coincides most clearly with the Great Depression. The levelling off in the 1970s in patent applications is also noteworthy.* The graph shows the effects of both World War I (1918) and World War II (1941-2). Overall, and despite the relatively weaker evidence for the 1875-84 depression (which was in any case relatively insignificant in the USA compared with Europe), Figure 3.4 tends to support the view that deep depressions lead to a serious reduction in inventive activity. It does not of course provide an answer to the question of the possible fluctuations in *radical* or *fundamental* inventions.

*The sudden decline in 1979 in the number of patents granted does not signal any dramatic fall-off in technological activity, but merely a shortage of resources for printing in the USA Patent Office. The general trend in the number of patents applied for and granted in the USA is, however, somewhat inflated for the most recent period, in particular from 1960 onwards, because of the very rapid growth of foreign patenting in the USA, primarily as a result of the increased international diffusion of technology.

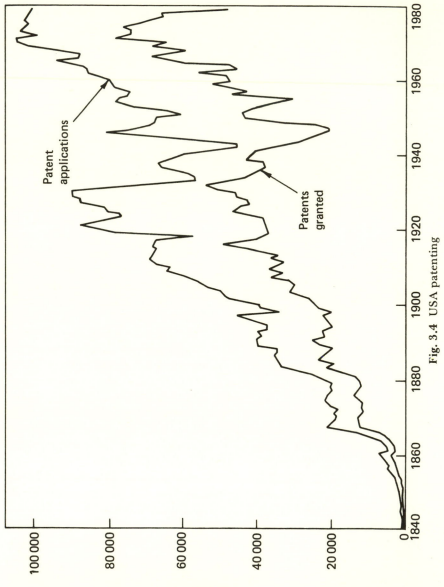

Fig. 3.4 USA patenting

The advantage of patent statistics is that they provide readily available information over a rather long period which can be easily classified by year and is not affected by changes in relative prices. A more important advantage is however that they have been collected and examined over all these years by the same official agency, generally speaking a Patent Office. Moreover, from 1734 onwards all patents issued in the UK have a specification describing the invention in full. It is on the basis of these specifications that an attempt was made by Baker (1976) of the British Library to select 'significant' inventions. Baker's method consisted of identifying first 'significant' subjects, for which the most important inventions, in terms of the major patents, could then be identified. These ranged from the Addressograph to zip fasteners, and constitute, in our view, a rather coherent and comprehensive sample of the major inventions of the last two centuries. For a number of subjects, Baker also identified some specific subclasses; for example, in the case of radio valves and microwave tubes, the diode, triode, pentode, variable mu, klystron, cavity magnetron, and travelling wave tube subclasses were also considered. Including these subclasses, some 400 different major subjects were identified for which information, including the exact year, concerning the first patent on the subject, and what we would call the *master* and the *key* patents, was given. The master patent is defined as 'the first to be economically viable' (Baker 1976, p. 15), while key patents refer to the most important patents in relation to each specific subject. The total number of patents covered by Baker's analysis was about 1000 and covered the period from 1691 to 1971.

On the basis of this information, ten-year moving averages of the master patent data and the key patent data for the period from 1775 to 1965 are plotted in Figure 3.5. The evolution of the number of first patents was practically identical to the master patent evolution ($r = 99.7$ per cent) and has not been drawn here.

A number of points emerge from Figure 3.5. First, the evolution of the number of key patents is very similar to the evolution of the number of master patents ($r = 97.7$ per cent). The major difference relates to the most recent period, understandably a period for which it is rather difficult to judge what will become 'master' patents. The close correlation between the three patent concepts suggests however that the criteria for selecting key patents as used by Baker is relatively reliable. Secondly, despite the steady increase in the number of inventions over time, the data support the view that 'clustering' of basic inventions takes place, and that the evolution over time of such inventions as measured by numbers of key or

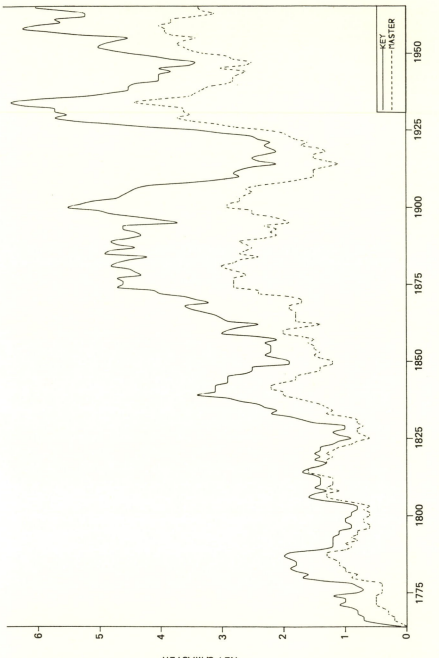

Fig. 3.5 Significant patents, 1764–1969 (ten-year moving averages)

master patents, is not characterized by a linear trend. Thirdly, it seems possible to identify a 'clustering' of major inventions in various phases of the long wave, including depressions (clusters in periods 1874-89 and 1928-36), prosperity phases (clusters in periods 1897-1903 and 1956-61), and war (1806-15). There does not therefore appear to be clear *prima facie* evidence that the observed clustering is unambiguously related to particular economic circumstances, whether favourable or adverse.

While these data are highly imperfect, they probably are among the best available and the trends they suggest are *grosso modo* inconsistent with the overall patent evolution as illustrated in Figure 3.4. This confirms the importance of examining data on 'significant' inventions and innovations rather than merely aggregate data.

Summing up a rather complicated discussion we conclude the following:

(i) Basic *inventions* do show a tendency to cluster at certain periods, including a big cluster in the early 1930s, but these clusters are apparently not systematically related to depressions.

(ii) Neither does the available evidence consistently support the Mensch–Kleinknecht theory of heavy clustering of basic *innovations* in periods of deep depression, although there is some evidence of a cluster in the late 1930s.

(iii) Nor does it support the acceleration hypothesis of reduced lead-times for innovations launched in deep depressions.

(iv) There is however evidence of a falling off in basic innovations at the tail-end of long booms.

(v) Firms tend to reduce both their R and D activities and their patenting during the more severe depressions.

But in rejecting the Mensch theory have we not cut the ground from under our own feet and destroyed also the argument in favour of Schumpeterian discontinuities in the innovation process? We do not think so but hope rather that we can now put the whole argument on a more secure foundation, which we attempt to develop in the following chapters.

4 'NEW TECHNOLOGY SYSTEMS': AN ALTERNATIVE APPROACH TO THE CLUSTERING OF INNOVATIONS AND THE GROWTH OF INDUSTRIES

In Chapter 3 we criticized Mensch's theory of clustering of innovations through depression-induced acceleration of the gestation period between basic inventions and basic innovations. In this chapter we shall attempt to develop an alternative theory which places greater emphasis on the role of scientific discoveries, on the technical and social inter-relatedness of 'families' of innovations and on the many follow-up innovations made during the diffusion period. We call these clusters 'new technology systems', because although they are associated with the rapid growth of one or more new industries, they often also have wider effects on other industries and services.

4.1 Clusters of innovations and the diffusion of innovations

We would expect to find a considerable variety of clustering processes and the phase of the long wave would in our view be only one of the influences on this clustering. Others would be new developments in fundamental science, particular breakthroughs in technology, wars and preparation for wars, marketing and organizational developments within the system of production and distribution, and finally the 'natural trajectory of technologies', discussed by Rosenberg (1976), Nelson and Winter (1977) and Dosi (1982). This does not mean abandoning the Schumpeterian notion of a reciprocal association between innovation and long-term economic fluctuations, but the nature of this relationship is more complex and untidy than a clustering of a large number of 'basic innovations' in particular decades every half century or so.

In enumerating and studying clusters of innovations it is essential to take on board Keirstead's (1948) point about innovations of 'limited adaptability' and innovations of 'wide adaptability' and his point that for the latter group the links between related innovations are scientific, technical *and* economic. We are interested in 'constellations' of innovations which have a relationship to each other and not just in the more or less accidental statistical grouping of the innovations of a particular year or decade. The important phenomenon to elucidate if we are to make progress in understanding the linkages between innovations and long waves is the birth, growth, maturity and decline of *industries and technologies*. The introduction of a major new technology into the economic system can take a matter of decades and affect many industries but the process has cyclical aspects which can give rise to long wave phenomena. In Chapter 5 we shall try to illustrate these points with respect to macro-molecular chemistry and the plastics industry, and in Chapter 6 we shall illustrate them with respect to electronics and its many applications.

In concentrating attention on the statistical aspects of enumerating 'basic' innovations and relating them to big depressions, Mensch has actually missed the main point of Schumpeter's theory on the reciprocal effects of innovation and the state of the economy. The macro-economic effects of any basic innovation are scarcely perceptible in the first few years and often for much longer. What matters in terms of major economic effects is not the date of the basic innovation (important though this may be for other purposes); what matters is the diffusion of this innovation — what Schumpeter vividly described as the 'swarming' process when imitators begin to realise the profitable potential of the new product or process and start to invest heavily. This swarming may not necessarily occur immediately after a basic innovation although it may do so if other conditions are favourable. In fact, it may often be delayed for a decade or more until profitability is clearly demonstrated or other facilitating basic and organisational innovations are made, or related social changes occur. Once swarming does start it has powerful multiplier effects in generating additional demands on the economy for capital goods (of new and old types), for materials, components, distribution facilities, and of course labour. This, in turn, induces a further wave of process and applications innovations. It is this combination of related and induced innovations which gives rise to expansionary effects in the economy as a whole.

We agree wholeheartedly with Rosenberg's point (1976, Chapter 11) that the diffusion process cannot be viewed as one of simple replication and carbon-copy imitation, but frequently involves a string of further innovations — small and large — as an increasing number of firms get involved and begin to learn new technology and strive to gain an edge over their competitors. None the less it is this diffusion, with or without further innovations and improvements, which alone can give rise to significant economy-wide effects on the pattern of investment and employment. Consequently, it will sometimes be the case that the basic innovations which have a major impact in a particular long upswing will actually first have been made in a different Kondratiev cycle altogether. This will apply *a fortiori* to the international diffusion of technology. Everyone who has studied detailed case histories of innovations is familiar with the long gestation periods and false starts which often occur. The controversy illustrated in Table 3.1 in the previous chapter clearly demonstrates this point.

From this standpoint the date of a particular basic innovation (or series of basic innovations) is less important than the interaction of a cluster of innovations or a social change which permits a market to grow rapidly or large amounts of capital to be raised and invested. Historians may argue for a long time about when to date 'railways' as a 'basic innovation' — whether it was 1829, 1825, 1817 or even earlier because of the railways in the mines. But the important phenomenon was the railway building boom in the second half of the century (although there were important national variations) and the construction of national railway networks, with their enormous demands on the iron and steel and engineering industries, on available investment resources, on construction labour, on engineering skills and the many additional innovations made in the course of this expansion.

Thus the set of innovations which are diffused and exploited during a Kondratiev upswing will not only be those of the immediately preceding depression, but will comprise some made earlier, some made during the depression and some during the recovery and upswing itself. This explains why we would attach rather less importance to the statistical clustering of basic innovations and much more to their linking together in new technological systems. However, depressions may help to bring about big changes in the social and political climate (as opposed to business behaviour in firms) and these in turn may generate conditions that are more favourable in the recovery phase, both for new basic innovations and for the swarming process around older basic innovations that may have been introduced at various times but are only able to flourish when the necessary social environment favours their rapid adoption. This may be occurring now in relation to the social and political conditions affecting telecommunications and information technology.

Mensch has been looking at the wrong 'swarms'. Surprisingly, when he speaks of the 'bandwagon' effect in relation to a cluster of basic innovations (1975, p. 193), he is apparently not talking about Schumpeter's 'one-sided rushes' of firms anxious to jump in and get above-average profits in a rapidly growing new branch of industry, he is talking about a disparate set of basic innovations. It is very hard to see in what sense the originally quite separate launch of helicopters, television, tetra-ethyl lead, titanium, etc. in the mid-1930s could constitute a 'bandwagon' in any normal meaning of the term. The swarms which matter in terms of their expansionary effects are the diffusion swarms *after* the basic innovations and the swarming effects associated with a set of inter-related basic innovations, some social and some technical, and concentrated very unevenly in specific sectors.

The bandwagon effect *is* extraordinarily important — in our view it is the main explanation of the upswings in the long waves. It is the steep part of the 'S-curves' characteristic of many diffusion processes, not the relatively flat piece of the curve which often follows the basic innovation for a few years. The bandwagon effect is a vivid metaphor and it relates to a rapid diffusion process which occurs when it becomes evident that the basic innovations can generate super-profits and may destroy older products and processes. The big-boom phase of the post-war Kondratiev could be described as the roughly simultaneous rolling of several new technology bandwagons; for example, the computer bandwagon, the television bandwagon, the transistor bandwagon, the drugs bandwagon and the plastics bandwagon were all rolling fast in the 1950s as well as some other bandwagons like consumer durables. These tend to harmonize partly because profit expectations and market expectations more generally will all tend to favour the expansion of new industries.

It follows from this that the discontinuities that most interest us are something more than statistical fluctuations in numbers of discrete basic innovations and more, too, than variations in the rate of diffusion of such innovations — although we think that these are certainly important. Even if it is granted that the S-curve pattern is characteristic of many innovation diffusion patterns, in principle there is no reason why these should be synchronized to generate major fluctuations in the economy as a whole. If the S-curves were randomly distributed for a fairly large number of discrete basic innovations, and if, too, the shape of the curves varied quite a lot, then there might be a series of ripple effects in the larger economy but there would not have to be big waves.

Big wave effects could arise either if some of these innovations were very large and with a long time span in their own right (e.g., railways) and/or if some of them were interdependent and interconnected for technological and social reasons or if general economic conditions favoured their simultaneous growth. Thus we are interested in what we shall call 'new technology systems' rather than hap-hazard bunches of discrete 'basic innovations'. From this standpoint, which we believe was essentially that of Schumpeter, the 'clusters' of innovations are associated with a technological web, with the growth of new industries and services involving distinct new group-ings of firms with their own 'subculture' and distinct technology, and with new patterns of consumer behaviour. Schumpeter spoke of the first Kondratiev as based on a cluster of textile innovations and the widespread applications of steam power in manufacturing, the second as the railway and steel Kondratiev, and the third as based on electricity, the internal combustion engine and the chemical industry. In later chapters we shall describe some of the main tech-nological systems associated with the fourth (post-war) Kondratiev.

4.2 Diffusion theory

Here, it is important to note an important development in 'diffusion' theory. During the 1960s a 'standard model' of diffusion of innova-tions was developed by Mansfield (1961) and others, emphasising the role of profitability for potential adopters, the scale of invest-ment required for adoption and the learning process within the population of potential adopters as the key determinants of the diffusion or adoption process. This model, although very useful for many purposes, neglects changes in the environment during the process of diffusion and changes in the innovation itself during that process. In the twenty years since the development of this model a good deal of empirical work on diffusion of innovations has provided a better basis for generalization in this previously neglected area.

In a seminal paper Metcalfe (1981) has pointed out that the role of profitability for *suppliers* (as opposed to adopters) is ignored in the 'standard' model, and so, too, is the influence of secondary innovations affecting profitability both for suppliers and adopters.

Gold (1981) and Davies (1979) . . . have argued that observed diffusion paths primarily reflect changes in the innovation and adoption environment rather than a process of learning within a static situation. As Gold observes, the stand-ard diffusion model rests on the implicit static assumption that the diffusion levels reached in later years also represent active adoption prospects during earlier years. . . .

The demonstration of profitability for suppliers is just as import-ant as the demonstration of profitability for adopters, as only this will normally induce the expansion of new capacity and skills neces-sary to sustain a rapid adoption process. Emphasis on the role of 'change agents' is in this respect more realistic than the predominantly demand-oriented standard economic model.

In Schumpeter's model the profits realised by innovators are the decisive impulse to surges of growth, acting as a signal to the swarms of imitators. The fact that one or a few innovators have made exceptionally large profits does not mean, of course, that all the imitators will do so. It is enough that they hope to, or even that they hope to make a fraction of them. Nobody else made such profits from nylon as Du Pont or such profits from computers as IBM. Indeed some would-be imitators made losses. This is an essential part of the Schumpeterian analysis. As the bandwagon begins to roll profits are competed away and some people fall off the wagon. Schumpeter himself stressed that changing profit expectations during the growth of an industry are a major determinant for the sigmoid pattern of growth. As new capacity is expanded, at some point (varying with the product in question), growth will begin to slow down. Market saturation and the tendency for technical advance to approach limits, as well as the competitive effects of swarming and changing costs of inputs, may all tend to reduce the level of profitability and with it the attractions of further investment. Exceptionally, this process of maturation may take only a few years, but more typically it will take several decades and sometimes longer still. The effects of saturation of consumer demand are discussed further in Chapter 7; here we pursue the argument mainly in relation to supply.

In his own new model of the diffusion process, Metcalfe is concerned to redress the balance between the demand side and the supply side. He incorporates into his model the profitability of the innovation for suppliers (as well as for adopters), and the growth of capacity on the supply side, which are ignored in the earlier standard adoption models. In this, and other, ways he is forging a link between modern diffusion research and the earlier work of Kuznets (1930), Burns (1934) and Schumpeter (1939) on the patterns of industrial growth. The emphasis shifts from the model of the isolated diffusion path of each discrete unchanging innovation to a succession of related innovations associated with the emergence, growth, maturity and (sometimes) decline of an industry. This earlier research now being re-discovered had also stressed the role of profitability for suppliers and had pointed to a combination of factors which would ultimately tend to retard growth and erode profit margins, thus giving rise to cyclical patterns extending over decades rather than years. Among these factors were market saturation, the extent to which the innovation was competitive with existing product lines (or complementary to them), inelasticities in the supply of inputs — some of which might be prolonged and tend to raise costs — and the pattern of induced secondary innovation. The impact of incremental technical improvements would tend to diminish gradually in accordance with Wolff's Law.*

4.3 Diffusion, process innovation and social change

From what has been said it is evident that we would expect radical or basic innovations to be spread rather more at random over various phases of the long wave (Figure 3.1) than either Mensch or van Duijn. Nevertheless, so far as a particular industry is concerned we would accept the general point made by them — and more systematically by Abernathy and Utterback (1978 and 1979) and Sahal (1980) — that there is a tendency for the pattern of innovation to shift over time through a life cycle from products to processes (strongly associated with economies of scale as an industry matures) and ultimately to process and product improvements of a relatively minor variety.

So far we have tried to insist that what matters from the standpoint of large-scale economic fluctuations is not so much the date of a particular basic innovation as a constellation of circumstances favourable to the exceptionally rapid growth of one or more new industries, each involving the combination of a number of related inventions, innovations and economic and social changes. We would now insist furthermore on the vital importance of Schumpeter's point about managerial and organizational innovations. These may often be just as important as the technical changes for the growth of an industry or technology. Thus, for example, many applications of the steam engine required the reorganisation of production on a factory basis, which was an extremely painful and difficult social change at the time. Some of the applications of robotics and microelectronics, both in manufacturing and service industries, may similarly require extensive social changes. These are likely to be spread over decades rather than years.

The adoption of many new electronic information systems, such as tele-shopping and tele-banking, will depend on institutional and

*Wolff's Law: Wolff was a German economist who in 1912 published four 'laws of retardation of progress'. Essentially he argued that the scope for improvement in any technology is limited, and that the cost of incremental improvement increases as the technology approaches its long run performance level. Widely referred to in the retardation literature of the 1930s (see, e.g., S. Kuznets, Secular Movements in Production and Prices, Houghton Miflin, 1930, Chapter 2).

legal changes, standards and other major government decisions in regional telecommunications investment. The social and political climate in particular countries at particular times may or may not be especially hospitable to these types of social and organizational change. The capacity for social innovation is very variable and in addition to the capacity to generate and launch a particular group of technical innovations this must surely be one of the main reasons for the changing locus of technological leadership in the various long waves, to which Ray (1980) has so rightly drawn attention and which we take up in Chapter 9.

The capacity for social change and innovation was rather high in England in the eighteenth and nineteenth centuries (although bought at a price in working class suffering whose aftermath is still with us). The development of the first major railway network in the world involved not just the invention and innovation of the railway locomotive (which had occurred long before) but a whole series of other inventions and innovations affecting the mechanical engineering industries and the iron and steel industries, as well as necessitating changes in the capital market, in legislation and in the training of a skilled labour force. It seems that the capacity of the Japanese to introduce applications innovations and the related social and organizational changes is (for very different reasons) rather high in relation to micro-electronics, robotics and biotechnologies. This might give them a leading role in any new upswing of the world economy even if they have not made the majority of basic innovations which may be associated with that upsurge.

A good example of the points we are trying to make is the automobile industry and Klein (1977) has a nice chapter in his book on *Dynamic Economics* which illustrates most of them. The USA did not lead Europe in the earlier stages of the innovation of the internal combustion engine or the passenger car — rather the reverse, in fact. Indeed it was by no means clear that the internal combustion engine would be preferred to the steam engine or electric engines. The basic innovation for all three had been made well before the turn of the century although the actual dates are still a matter of controversy. Klein points out that in 1900 steam and electric vehicles accounted for 'about three fourths of the four thousand automobiles estimated to have been produced by fifty-seven American firms' (1977, p. 91). The decisive step which the US firms made (as a result of the competitive pressures within the industry) was to reduce the cost of manufacture of the gasoline engine car by more than 50 per cent within a few years. The price of a Model T fell from $850 in 1908 to $360 in 1916, sales increased by a factor of 50, market share increased from 10 per cent in 1909 to 60 per cent in 1921, profits on net worth were sometimes as high as 300 per cent per annum and the US attained a dominant position in world export markets. This was indeed 'fast history' analogous to the tempestuous growth of the semi-conductor industry half a century later — with its similar drastic price reductions, rapid changes in market shares, sudden profits for innovating firms and world export hegemony for the leading country until imitators caught up.

The 'basic innovation' which enabled Ford to achieve these dramatic results was, of course, assembly line production. The date of this innovation was right in the middle of the long-wave upswing, not in the earlier depression. In one sense it was a purely organizational innovation, but it both entailed and stimulated a great deal of technical innovation:

. . . once the organisational change was made, the automobile firms found many opportunities for developing more efficient machines by making them more automatic. For example, replacement of the vertical turret lathe by a more automatic horizontal lathe doubled output per worker. Or to take a more spectacular example, an automatic machine for making cam-shafts increased output per worker by a factor of ten, and literally dozens of cases can be found in which better machines permitted output per worker to increase by a factor of between two and ten.

Incidentally, although strongly Schumpeterian in the main thrust of his argument, Klein casts some doubt on the 'heroic' entrepreneur idea so far as it applied to Ford, quoting the comment of Nevins and Hill (1954):

It is clear that the impression given in *My Life and Work* that the key ideas of mass production percolated from the top of the factory downwards is erroneous: rather seminal ideas moved from the bottom upwards. To be sure, Ford took a special interest in the magneto assembly. But elsewhere able employees like Gregory, Klann and Purdy made important suggestions, Sorensen and others helped them work out, while Ford gave encouragement and counsel. The largest single role in developing the new system, however, was played by the university-trained thinker [Avery] so recently brought in from his school-room.

As we shall see in Chapters 5 and 6, the linkages with university research were very much stronger in most of the technologies associated with the fourth Kondratiev, but once the industrial application of a new technology begins to develop, to a considerable extent it has its own momentum.

4.4 'Natural trajectories' of technologies

The expression 'natural trajectory' has been coined to describe this process of cumulative exploitation of new ideas. In their paper 'In search of useful theory of innovation' Nelson and Winter (1977) distinguish various types of natural trajectory, including some which are specific to a particular industry or product and some which are of very great general importance such as mechanization. To the best of our knowledge these authors have not attempted directly to relate these ideas to 'long waves' in the economy as a whole but they do point out that:

. . . there is no reason to believe (and many reasons to doubt) that the powerful general trajectories of one era are the powerful ones of the next. For example, it seems apparent that in the 20th century two widely used natural trajectories opened up (and later variegated) that were not available earlier: the exploitation of understanding of electricity and the resulting creation and improvement of electrical and later electronic components, and similar developments regarding chemical technologies . . . it is apparent that industries differ significantly in the extent to which they can exploit the prevailing general natural trajectories, and these differences influence the rise and fall of different industries and technologies.

Such general processes of technical change as mechanization, electrification or automation obviously continue over a century or more, but as Coombs (1981) has suggested the main thrust of their application may be changing, for example, from transfer applications to control applications. Nelson and Winter point out that these very general trajectories of technology are associated strongly with the exploitation of economies of scale. One could therefore expect that, as an industry enters upon a period of rapid expansion, fairly intense efforts will be made to promote innovations which facilitate the attainment of these economies — as was found in the study of chemical process plant in the 1960s (Freeman *et al.* 1968) and in the semi-conductor industry in the 1950s and 1960s (Chapters 5 and 6). The example of the automobile industry in the USA also illustrates the point rather nicely. The history of these 'process trajectory' innovations is likely to be imperfectly documented by comparison with product innovations that are better publicized and are, in any case, easily accessible to the general public. This is an important source of bias in innovation statistics. The computer may appear only once in a list of dates of innovation but its on-line application in a great variety of process control systems may not appear at all.

Secondly, although the search for factor-saving technical change is a constant feature of the innovative efforts of many industries, it is likely to be particularly intense in certain circumstances. One obvious case is the contemporary emphasis on energy-saving technical innovations, apparently increasingly effective in the case of oil, although with a time lag which is directly relevant to the discussion. Another case is the desire to save on imported natural materials in the event of war or threat of war. In the case of labour costs it seems possible that the pressure for labour-saving technical change will be at its most intense during periods of labour shortages and a steep relative rise in labour costs (i.e., at peak periods of prosperity). However, because of the time lags involved in any such change of emphasis and the independent application of such innovations, the actual shedding of labour may occur somewhat later.

Thus, although the empirical evidence is still inadequate, major general natural trajectories, such as mechanization, electrification or automation may relate to long waves along the following lines. First, a basic innovation — the steam engine, the electric motor or the computer — and a small cluster of related basic innovations create possibilities of revolutionary changes in the methods of production in a wide variety of industries and services. Firms which are producing and designing the new types of capital goods experience a surge of extremely rapid growth and many new firms enter these industries (Coombs 1981). Secondly, the other fast-growing industries of the upswing of the long wave, which are in a position to do so, make use of these natural trajectories to exploit economies of scale and to achieve very high rates of productivity increase. The scale of their investment permits large-scale introduction of new technology. Here, we could expect further clusters of process innovations and instrument innovations, some specific to particular industries and others of more general importance. Thirdly, as the sustained expansion generates labour shortages and inflationary pressures on labour costs, profitability tends to decline and there is an increasing induced demand for labour-saving technical innovation throughout the economy, exploiting the potentiality of the most recent general natural trajectories. Applications here, however, may often be small scale and piece meal because of existing structures and traditions, and the lack of resources and skills. Some old-established but declining industries may succeed in achieving big productivity increases through rationalization and structural adaptation, accompanied by more investment, but the full exploitation of the potentiality of major new technologies in these sectors may often have to await another major expansionary phase.

In Chapters 6, 7 and 8 we attempt to show that such a hypothesis is supported — not only by the evidence of developments in the electronics industry, but also by the evidence of changing capital intensity and differential productivity increases in manufacturing more generally during the 1960s and 1970s — as well as by the shift from capacity extension type investment to rationalization. On these lines of argument we could expect further clusters of process innovation to occur all through the prosperity and stagnation phases of the long wave, although many of them would be hard to identify since they would constitute special applications of computers and a variety of instruments.

4.5 New technology, employment and long waves

We now attempt to draw together this rather extended discussion and relate it to the question of employment (Table 4.1). In the major boom periods new industries and technological systems tend to generate a great deal of new employment, as the form that expansion takes is the installation of completely new capacity and the building up of associated capital goods industries. Since the technology is still in a relatively fluid state and standardized special plant and machinery is not yet available, the new factories and plants are often fairly labour-intensive. New small firms may also play an important role among the new entrants and they tend to have a lower than average capital intensity.

However, as a new industry or technology matures, several factors are interacting to reduce the employment generated per unit of investment. Economies of scale become increasingly important and these work in combination with technical changes and organizational changes associated with increasing standardization. The profits of innovation are diminished both by competition and by the pressures on input costs, especially labour costs. A process of concentration tends to occur and competition forces increasing attention to the problem of cost-reducing technical change. This tendency plays an important part in the cyclical movement from boom to recession (or stagflation) and from stagflation to depression.

Hicks (1974), Scitovsky (1978) and others have suggested a plausible mechanism by which inflationary labour cost pressures could be generated and transmitted during the boom and stagflation phases of a long wave. There will be persistent skill shortages during the rapid growth of new technological systems as the generation and spread of a new technology by definition calls for the deployment of skills which have not hitherto existed or only partially

Table 4.1 A simplified schematic representation of new technological systems

	Previous Kondratiev	Recovery and boom	'Main carrier' Kondratiev	
			Stagflation	Depression
Research, invention	Basic inventions and basic science coupled to technical exploitation. Key patents, many prototypes, many basic innovations.	Intensive applied R and D for new products and applications, and for back-up to trouble shooting from production experience. Families of related basic innovations.	Continuing high levels of research and inventive activity with emphasis shifting to cost-saving. Basic process as well as improvement inventions are sought.	R and D-investment becomes less attractive. Despite the fact that firms try to maintain their level of research it becomes increasingly difficult to do so with the slackening of their sales. At the same time the volume of sales required to amortize the cost of R and D is steadily increasing. Basic process innovations still attractive to management but may meet with social resistance.
Design	Imaginative leaps. Rapid changes. No standardization, competing design philosophies. Some disasters.	Still big new developments but increasing role of standardization and regulation.	Technical change still rapid but increasing emphasis on cost and standard components.	Routine 'model' type changes and minor improvements of cumulative importance.
Production	One-off experimental and moving to small batch. Close link with R and D and design. Negligible scale economies.	Move to larger batches and where applicable flow processes and mass production. Economies of scale begin to be important.	Major economies of scale affecting labour and capital but especially labour. Larger firms.	The slowdown in output and productivity growth leads to over-production and excess capacity in some of the modern industries. These structural problems are 'cumulative and self-reinforcing', with repercussions for the economy at large, and lead to a further decline of economic activity.
Investment	High risk speculative, small scale. Some inventor-entrepreneurs. Some	Bunching of heavy investment in build-up of new capacity. Band-wagon	Initially continuing heavy investment but shifting to rational-	Relatively low levels of investment. Underutilization of the capital stock in some of the most modern

	large firms. Fairly labour-intensive. Problems of venture capital.	effects. Large and small firms attracted by high profits and new opportunities.	ization. Continuing rapid growth, but increasingly large sums required to finance R and D and rising capital costs. Rising capital intensity.	sectors of the economy; low profit margins and the general 'pessimistic mood' with regard to expectations lead entrepreneurs to be very (over) cautious in relation to new investment opportunities. Investment which will take place will be primarily directed towards rationalization. Search for new investment opportunities abroad.
Market structure and demand	Innovator monopolies. Strong consumer resistance and ignorance. Some new small firms to promote basic innovations.	Intense technological competition for better design and performance. Falling prices. Big fashion effects. Many new entrants in early build-up.	Growing concentration. Intense technological competition and some price competition. Strong pressure to export and exploit scale economies.	Even stronger trend to oligopoly or monopoly structure. Bankruptcies and mergers.
Labour	Small-scale employment generating effects. High proportion of skilled labour, engineers and technicians. Training and learning on the job and in R and D	Major employment generating effects as production expands. New training and education facilities set up and expand rapidly. New skills in short supply. Rapid increase in pay.	Employment growth slows down, and as capital intensity rises, some jobs become increasingly routine.	Employment growth comes to a halt. Unemployment rising. In addition to the continuing labour displacement effects of rationalisation investments, employment suffers (in the first instance) from the general recessional and depressional tendencies in the economy at large.
Employment effects on other industries and services.	Negligible, but imaginative engineers, managers and inventors are thinking about them and planning and investing accordingly.	Substantial secondary employment, mainly employment generating but gradually swinging to displacement.	Labour displacement effects, as new technology now firmly established and strongly cost-reducing.	Continuing labour displacement as new technology penetrates remaining industries and services.

existed. Lags in the education and training system are often prolonged, and salaries of groups such as electronic engineers, computer
programmers and systems analysts and their equivalents in other
new technological systems will tend to increase more rapidly than
average. Firms in the most rapidly expanding sectors will also tend
to pay above average for 'conventional' skills in order to attract
good-quality labour for their new plants. It is not difficult to see
how, in a period of full employment and with the assistance of
'comparability' transmission mechanisms and strong union bargaining
power (or without it), inflationary labour cost pressures could be
generated (and sustained because of lag factors) throughout a period
of stagflation.

Here the Marxist tradition in economics and particularly Mandel
(1972) have made an important contribution to the debate in their
emphasis on the significance of the general rate of profit and the
tendencies which may lead this rate to fall. However, Mandel (and
Kuznets) were incorrect in asserting that Schumpeter's theory of
the long waves was dependent on a *deus ex machina* of waves of
entrepreneurial energy and ignored the importance of profit. The
role of profit was built into Schumpeter's theoretical system by
definition and he saw innovation as the source of a new range of
profit opportunities. However, the 'competing away' of profits
during the diffusion of new technologies may well be complemented
by the process which Mandel emphasizes — the tendency for the rate
of profit to fall throughout the system for reasons associated with
the growth of capital intensity and because of greater bargaining
strength of labour following long periods of full employment, and
related inflationary pressures and social changes.

In the fourth Kondratiev, in particular, it may be that the firms
involved in promoting the new wave of technological systems reflect
the more oligopolistic nature of capitalism generally and have been
better able to delay the 'competing away' of innovation profits.
Schumpeter already commented on this tendency in the 1920s,
and the strong inflationary pressures during the recession of the
fourth Kondratiev may be explained to some extent by a combination of this tendency and the stronger pressures on production costs,
including energy and material costs. This point is taken up in relation
to oligopoly and prices in Chapter 7: the extent to which the competing away of profits may be delayed varies from industry to
industry. In the drug industry the strength of patents was one
factor which led to much greater success in maintaining higher prices
and profits, than for example in the electronics industry.

As a result of oligopolistic structures and many other rigidities in the system, the period of 'stagflation' is likely to combine many paradoxical features (Kristensen 1981). In particular, surplus capacity in older declining industries (and even some new ones) is likely to persist side by side with acute shortages in other branches. Similarly higher rates of unemployment will exist side by side with acute skill shortages in some of the newer technological systems. There are other types of shortage in periods of boom and stagflation that can fuel inflationary pressures, but what interests us here is primarily the effect of such shortages and the associated cost pressures on the direction of technical change.

The combined effect of these cost pressures and the concentration, scale and standardization effects already mentioned will be to focus technical effort increasingly on cost-reducing process innovations rather than new products. The patterns of investment and R and D during a stagflation period will both tend to shift accordingly — the proportions of rationalization and replacement investment will tend to rise and new capacity investment to diminish; more investment will go into machinery and less into new buildings. In this period, too, there may be increasing pressure for 'improvement innovations' and 'product differentiation'. We would not expect basic innovations to stop but they may tend to slow down (Figure 3.1) and there will be greater pressures favouring basic process and other cost-saving innovations. Walsh *et al.* (1979) has shown that process patents rose much more rapidly than product patents in the plastics industry in the 1950s and 1960s (see Chapter 5).

Finally, the descent from stagnation to depression may be explained by 'overshoot' arising from efforts to combat inflationary pressures but also by the successive weakening of the expansionary elements within the long wave. In particular, the shift in the main thrust of investment and technical development will mean that the investment required to generate further increments of new employment will be rising, so that even when investment is expanded during the successive Juglar cycle upswings as a result of Keynesian or other stimuli, it has less effect in expanding employment, unless it is directed to areas of very low capital intensity such as various government and administrative services.

At the same time, some of the 'S-curves' of rapid expansion of new products and technologies will have begun to level off for a variety of other reasons, for example, saturation of economic demand which is discussed in Chapter 7. The shape of these logistic curves will, in any case, vary a great deal and for some the rapid wave of expansion will be sustained for much longer than others. It is clear, for example, that in the case of synthetic materials much of the expansionary impetus was already lost in the 1970s as substitution effects were lessening (see Chapter 5). The rapid expansion of entirely new technological systems to compensate for these effects (even if the basic innovations have already been made), may be delayed by a variety of factors — rigidities, labour shortages, capital shortages, etc., as well as by the general business climate in a period of stagflation. A depression should not be necessary to generate a revival of growth and the task of intelligent economic and social policy is to find the way to stimulate a new flow of desirable combinations of technical innovations and social changes to prevent prolonged depression. This is the subject of our final chapter.

4.6 Conclusions

To draw together a rather extended and tentative set of arguments we must first note that the upswing of the long waves involves a simultaneous or near-simultaneous explosive burst of growth of one or several major new industries and technologies. Such a take-off becomes possible on the basis of a favourable combination of circumstances. These include the previous successful realization (at whatever dates) of some earlier basic innovations, such as the car and various electrical innovations in the 1880s or the computer, television and a family of synthetic materials in the 1930s and 1940s. This demonstration effect has to be on a sufficient scale to trigger Schumpeter's swarming process, with Rosenberg's caveat that diffusion is not simple replication but involves a further set of related innovations. As the new industries grow they generate a further set of process innovations associated especially with the exploitation of economies of scale and drawing upon the 'natural trajectories' of technology available at the time (mechanization, automation, computerization etc.). The very success of the long-wave boom and its initially rather labour-intensive character (particularly obvious in the early days of railways, cars and electronics respectively) generates strong demand for labour.

This is reinforced by all the secondary and multiplier effects of the expansion on the economy as a whole. It may well be, as Forrester (1981), Tinbergen (1981) and Mandel (1972) all suggest, that there are additional long-wave-generating factors involved in the delayed response of the capital goods sector to the rapid expansion of demand and the need to divert part of the increased output to the expansion of the capital goods sector itself.

In any event, at the peak of the boom the demand for labour

becomes so strong (both in the period before World War I and in the 1950s and 1960s) as to stimulate big flows of immigration into the leading industrial countries and to facilitate the entry of new groups into the labour force. This big change in the labour market, (with or without the pressures of trade unions) increasingly generates cost inflationary pressures, however. These are reinforced by Hicks–Scitovsky mechanisms of comparability claims with the wage and salary increases attained in the leading (technologically advanced) industries. The effect of these, and other, shortages is the erosion of profit margins as well as a shift in the emphasis of investment away from simple capacity expansion to rationalization and cost-saving innovation.

Both the Marxists and the neo-classicals are probably right in believing that the change in general levels of profitability is an important factor in the behaviour of the system, both at the upper and lower turning points of the long wave. At the upper turning point it stimulates a search for labour-saving technical change and improvement à la Mensch. Although this may lead to intense efforts at productivity improvement, such efforts are likely nevertheless to be actually less effective than during the expansionary phase of the long wave. Whereas economies of scale in the newly expanding industries were a major source of productivity gain in the upswing, the opposite effect is engendered through below-capacity working in the downswing. Moreover, job insecurity and the rising level of unemployment now conspire to reduce the co-operation of the labour force in the implementation of the technical innovations, which may depend for their success on wholesale reorganization and structural change with many plant closures.

Some basic innovations may continue to occur during stagnation and depression, but the swarming process will be inhibited by the low general levels of profitability and business expectations. Unlike Mensch, we would expect the depth of the depression to inhibit or delay basic innovations to some extent. Because of the diminishing flexibility of market economies generally, and the important role of the public sector in R and D and innovation policies, the role of public policy is crucial. But some exogenous stimulus to the system, which helps to restore the general level of profitability and improve economic expectations generally, may also be necessary. It would be a social disaster if either the prolongation of mass unemployment or repressive measures were used (as in the 1930s) as means of attempting to restore a higher level of profit expectations. We return to this point in Chapter 10.

5 DEMAND AND TECHNICAL INNOVATION IN THE GROWTH CYCLE OF THE SYNTHETIC MATERIALS INDUSTRY

It has been argued in Chapter 4 that the swarming process around an initial cluster of technically inter-related basic innovations gives an impetus to expansionary waves in the economy as a whole, even though these basic innovations may not necessarily occur during an immediately preceding deep depression. Once the swarming process is under way then a large number of secondary and induced inventions and innovations will follow, giving rise to the kind of pattern which Schmookler analysed. In this chapter we discuss the trend of basic and secondary invention and innovation in relation to the plastics industry and other sectors of the chemical industry.

5.1 The trend of basic inventions and secondary inventions and the growth of the synthetic material industry

As we have seen in Chapter 2, the notion that demand alone might largely determine the rate and direction of inventive activity and of innovation has been given some credence through Schmookler's work, and we have contrasted his 'demand-led' theory of invention with that of Schumpeter. In this chapter we shall draw on the work of Walsh *et al.* (1979), who attempted to test Schmookler's theory in four sectors of the chemical industry. We shall also draw on two earlier research projects carried out at the National Institute of Economic and Social Research and at SPRU (Freeman *et al.* 1963, Freeman *et al.* 1968) on invention and innovation in the plastics industry and chemical process plant respectively. We differ from Schmookler in not accepting that aggregate trends in patent numbers are a satisfactory substitute for a more specific investigation of the relative significance and qualitative importance of various discoveries, inventions and innovations. However hard it may be to measure, there is a real and important distinction between 'basic' and incremental innovations and inventions. To take a specific example: in the drug industry the category of 'new chemical entities' is an important one and the fact that this indicator declined in the 1960s whilst aggregate patent numbers continued to rise was of major significance for the industry's development.

Schmookler (1966) was aware of the strength of this counter-argument and was at pains to justify his view that the trend of aggregate patent numbers was the really significant indicator because technical change was a continuum in which 'important' inventions were not really so important as they sometimes seemed. Nevertheless, his own evidence on 'important inventions' compared with aggregated patent numbers showed that in all the four major industries which he investigated the 'important invention' cluster led the aggregate patent indicator at the outset (p. 81), as indeed common-sense would suggest. His technique of using the time trend of aggregate patent statistics is also of course very useful, although (as he recognized), there are some important qualifications to be kept in mind in using *any* patent statistics as a measure of inventions, especially the distinction between patentable and non-patentable technical advances and variations in the propensity to patent. But once a new industry or technology has begun to take off on the basis of a cluster of related basic inventions and innovations, then aggregate patent numbers can provide very useful indicators of the strength and direction of secondary and induced inventive activity over time. Even if new basic inventions might be counted in the totals, their numbers would be so small relatively that they would barely affect the aggregate trend.

In Figure 5.1 we have plotted the ten-year moving averages of the number of major discoveries, inventions and innovations in the plastics industry over the period 1825 to 1963 as identified by J. H. Dubois from the *SPE Journal* and *Plastics World* staffs (1967). To complete the picture we also added data in relation to the number of patents and total US production of plastics (in volume) as given by Walsh *et al.* (1979). Despite the difficulties of uniquely categorizing the advances listed by Dubois into inventions and innovations, it appears that these data do show a clustering or bunching of key inventions and innovations and, with a clear time lag, its impact on overall plastics patenting and actual production.

This supports the notion of bunches of basic inventions and innovations leading to the take-off of new industries and a further set of basic and secondary inventions and innovations, especially in the process field as the industries grow rapidly. It does not, however, demonstrate any direct connection between this process and the 'trigger' of depression. Independent corroboration of this analysis comes from Hufbauer (1966), who carefully documented the dates of all the major product innovations in each country. The list below summarises his analysis of the numbers of product innovations in each decade:

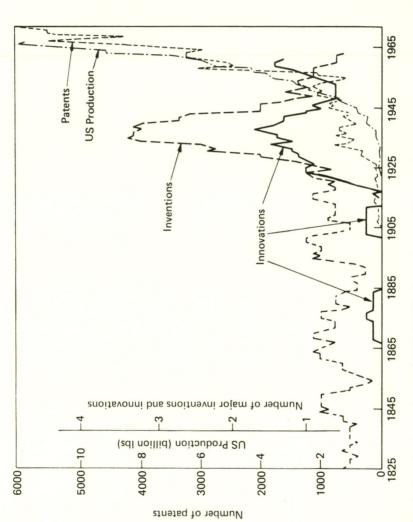

Fig. 5.1 The birth of the plastics industry (1825–1970). Inventions and innovations: ten-year moving averages (patents in actual numbers, USA production in lbs).

Before 1900	4
1900–1909	4
1910–1919	3
1920–1929	5
1930–1939	18
1940–1949	15
1950–1959	10
After 1959*	1

The SPRU project on the chemical industry (Walsh *et al.* 1979) also used long time series on patents, investment and production to investigate four subsectors of the chemical industry — plastics (Figure 5.2), dyestuffs, drugs and petro-chemical intermediates — and made use of an additional series — scientific papers — not used by Schmookler. The intention was to test Schmookler's hypothesis in an industry more closely related to basic science than the four which he studied (railways, paper, oil refining and agriculture) and for a more recent period.

Of particular interest was the evidence that, in some cases, an upsurge of inventive activity actually preceded a later upsurge of investment and sales. This was apparently the case with drugs in the 1930s and with both dyestuffs and plastics in the period 1925-35, although there are great difficulties in getting satisfactory German production and investment statistics for the plastics industry in the 1920s.

One-explanation of these partly 'counter-Schmookler' developments of the 1920s and 1930s is in terms of the German drive for national self-sufficiency which, prior to World War I and World War II, was designed to overcome dependence on imported natural materials of all kinds. This was to some extent independent of the short-term state of demand. However, this implies a considerable modification of the Schmookler model. This model attempted to explain the trend of invention as an activity which *followed* directly on the expansion or contraction of the market, rather than an activity which attempted to *anticipate* future changes in the market, or to create entirely new markets on the basis of scientific discoveries.

This weaker and more acceptable version of the Schmookler theory is less deterministic since the *anticipated* future developments are necessarily more uncertain and the interplay of new science, new technology and conjectured future markets becomes far more the realm of imaginative entrepreneurship. We have already argued in Chapter 2 that Schumpeter's models seem to have more to offer

*Hufbauer's work did not cover the 1960s.

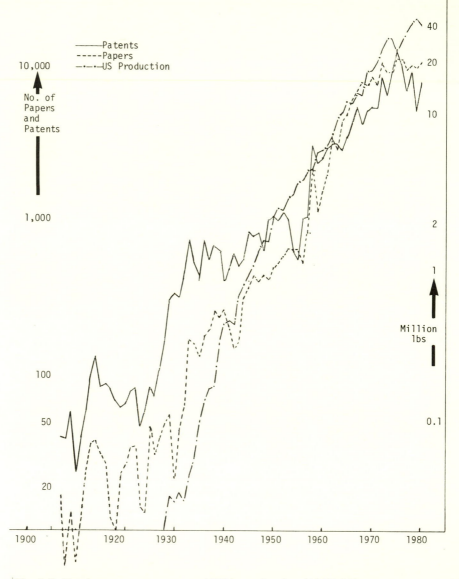

Fig. 5.2 Plastics patents, papers and USA production (1907–80).
Source: Walsh *et al.* (1979) and our updating.

here, but before we turn to them we first briefly consider what we shall designate as the 'Hessen' model, which is an even stronger form of the Schmookler model, extending its domain from inventions to basic scientific theories.

5.2 Basic science, invention and innovation

When Hessen (1931) and his Soviet colleagues first put forward their bold ideas at the International Congress on the History of Science in 1931, they caused something of a sensation, and ever since then their papers have continued to arouse controversy. Hessen was mainly concerned to demonstrate the profound influence of economic developments on even the most basic scientific ideas, and chose as his example the birth of Newtonian physics. Although he did not attempt to use the type of quantitative indicators which Schmookler used, if he was right in his main contention then we might expect the trend of publication of scientific papers to follow behind indicators of invention, since he was essentially expounding the view that technology leads and science follows.

At first glance the results of Walsh *et al.* (see Figure 5.2) appear to justify a Hessen model; there does seem to be some evidence in the case of plastics, dyestuffs and intermediate chemicals for a derived secondary demand for further scientific work following several years behind the original invention and innovation of new products or processes. Very large numbers of papers were published in the whole plastics field in the 1960s. It is difficult to resist the conclusion that the explosion of scientific papers concerned, for example, with acrylonitrile and caprolactam in the 1950s and 1960s were the consequence of the large-scale industrial use of these products for the plastics industry and the attempts to make new major and minor process innovations.

However, as in the analogous case of aggregate patent statistics, we have to consider the whole question of secondary publication in relation to the more seminal scientific papers and of original breakthroughs in scientific discoveries, which open up new fields for 'normal' science in Kuhn's (1962) sense and for 'gap-filling' (generating much larger numbers of derivative secondary publication chains) and for related inventive work. The *qualitative* analysis points strongly to the importance of these more fundamental developments as the starting points for later surges of publications, as well as inventions. There is an important parallel betweeen new scientific paradigms and new technological paradigms (Dosi 1982).

One of the earliest truly synthetic materials — bakelite — was invented and innovated by a Belgian chemistry professor (Baekeland) in 1909, but the surge of innovations in the 1930s was based on the fundamental research work of Staudinger in the early 1920s. This research was on long chain molecules and it led directly to the innovation of polystyrene and styrene rubber by I. G. Farben and indirectly to many other inventions and discoveries. It also provided a basis for the development of a new scientific discipline bordering on technology — polymer science.

It is important to note the peculiarly close relationship between 'science' and 'technology' in the plastics industry. An 'invention' is often virtually the same thing as a 'scientific advance' and may be the subject both of scientific papers and of patent applications. Price (1965), the historian of science who has emphasized most strongly the 'separateness' of science and technology, and who maintains that technologists generally do not seek recognition through publications but through artefacts and inventions, nevertheless makes an exception for chemistry and electronics. We shall discuss the case of electronics in Chapter 6, but here it is enough to note that there is substantial publication of papers by plastics 'technologists'.

As swarming began and the industry entered a period of rapid growth, work on new applications, on process improvement, on co-polymerization and on the modification of materials to meet the needs of specific users, entailed a great deal more scientific work on the properties and behaviour of many different long-chain molecules, as well as on emulsifiers, plasticizers, catalysts and other chemicals involved in their production. The chemical industry was also often keen to sponsor this research, whether in universities or in its own R and D laboratories. Taking into account the immense range of new applications which were developed in the war and post-war periods it is not surprising that the aggregate number of patents and publications soared upwards in the 1950s and 1960s, reaching a peak of over 10 000 papers annually in the early 1970s (Figure 5.2). The underlying scientific and technical links in synthetic material innovations were strongly emphasized by Hufbauer (1966), in his careful study of about sixty new materials and their international diffusion:

The similarity of nomenclature between various plastics, rubbers and man-made fibres indicates another important feature of the industry: materials with diverse economic uses are sometimes related chemically. It is this characteristic which justifies the designation of a synthetic materials 'industry'. . . . investigation of one high polymer often borders very closely on another. W. H. Carothers' discovery of nylon (both a plastic and a fibre) laid the groundwork for the discovery of polyester fibre by J. R. Winfield and J. R. Dickson in Britain. Likewise the I. G. Farben discovery of polystyrene plastic led naturally to the invention of styrene rubber and I. G. Farben's research on polymethyl methacrylate plastic opened the way to nitrile rubber which in turn led to acrylic fibre. . . . The technical bonds joining plastics, synthetic rubbers and man-made fibres are to some extent reflected in the concentration of innovation found among synthetic materials. Due to the chemical similarities of synthetics, one innovation by a given firm almost certainly increases the likelihood of the next. Aside from generating a pertinent body of high polymer research findings, a tradition of innovation attracts the brains and produces the financial successes requisite for its own propagation.

These comments illustrate very well the emergence of polymer science as a common 'sub-culture' for the industry, linking it closely with fundamental science in universities.

The German chemical industry was particularly remarkable for its close links with academic research. Ever since the German dye-stuffs firms began to promote product and process innovation on a systematic basis in the 1870s, they have deliberately cultivated close links between their own in-house R and D teams and universities — by consultancies, research grants and regular awards to university chemists. Staudinger was himself a consultant to I. G. Farben and so too was Ziegler, whose fundamental work on catalysis led to the major process innovation of the 1950s — the low pressure process for polyethylene — and indirectly to the major product innovation of the 1950s, polypropylene. The SPRU study showed in detail how the collaboration between Montecatini and Professor Natta of the Milan Polytechnic led to the innovation of polypropylene in 1958, preceded by a surge of both publications and patents by Natta's group and by Montecatini's own researchers (Figure 5.3). A second (and much bigger) upsurge of papers and patents followed later when many licensees and swarming imitators moved in to expand world production at an extremely rapid rate in the 1960s. This example shows the importance of detailed case studies of particular innovations to complement the aggregate analysis of papers and patents, since of course the aggregate growth of plastics secondary inventions and derivative scientific papers in the 1950s (Figure 5.2) completely swamped the small 'explosion' of publication and patenting associated with the innovation of polypropylene itself, shown in Figure 5.3.

However, even using the device of counting aggregate numbers of scientific papers as recorded in *Chemical Abstracts* there were instances in the chemical industry in which this indicator led both the indicators of inventive work and those of sales and production, the most notable being that of drugs in the 1930s (Walsh *et al.* 1979, Section 6).

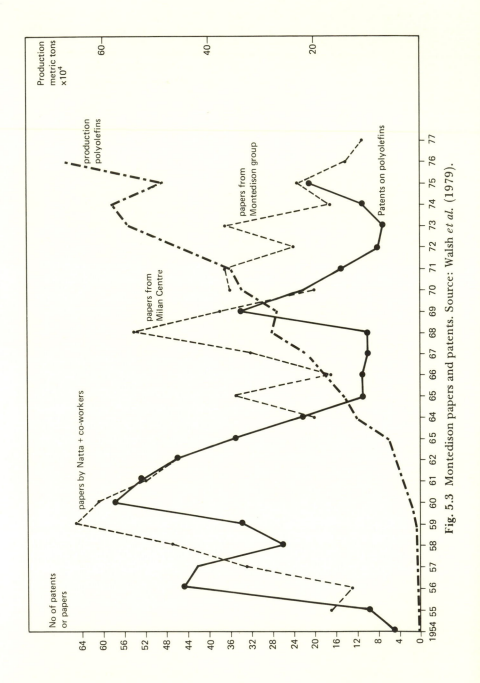

Fig. 5.3 Montedison papers and patents. Source: Walsh *et al.* (1979).

Independent evidence confirming the tendency for the indicators of basic research and of inventive activity sometimes to lead the upward surge in production in the early stages of the growth of an industry also comes from the work of Yamada on the Japanese plastics industry (1982). He has calculated 'life cycles' for research papers and patents based on the ratio of plastics papers and patented inventions to the total numbers of scientific publications and patents. Comparing these with a similar ratio for the Japanese plastics production life cycle, he shows that the various life cycle indicators for research and invention peak several years before the production cycle peak (Figure 5.4). His work is also of interest for several other reasons. It shows the 'institutional lag' in the formation of departments of polymer chemistry in Japanese universities and also shows that the life cycle of the Japanese plastics industry, although obviously differing in some respects from the European or American industries, followed a similar course in the post-war period.

The strength of Japanese research was such that PVC was actually produced on a small scale in the same year that it was launched in Germany (1931) and nylon too was produced in small quantities soon after the USA. Japanese researchers were contributing to the international literature in the field before the war and on a significant scale in the early post-war period (curve 1 in Figure 5.4). This is an important corrective to the commonly held view of a purely imitative response based entirely on the import of foreign technology. Yamada's work shows that independent research and invention were critical also in the Japanese case and that their life cycle peaks preceded the enormous expansion of Japanese production in the 1960s and 1970s.

5.3 Schumpeter's mark II model and the rise of the plastics industry

As we have seen in Chapter 2, the Schumpeter mark II model postulated that 'endogenous' science and technology, mainly within the R and D laboratories of the large firms, had increasingly substituted for the mechanism of the 'exogenous' inventor setting up in business. He did not, however, rule out the possibility of the mark I mechanism continuing to operate within a climate increasingly dominated by large-scale corporate R and D. The transition from mark I to mark II describes rather well what happened in the chemical industry. Originally in the nineteenth and early twentieth centuries, men like Perkin, Linde, Dow and Baekeland, coming often from an academic background, were able to make a fortune from their own discoveries

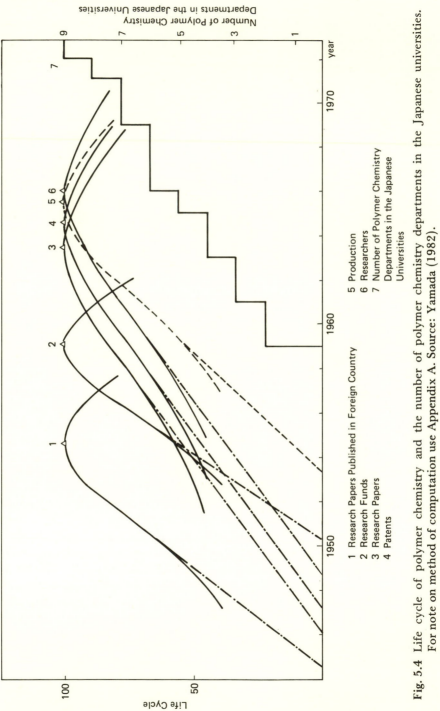

Fig. 5.4 Life cycle of polymer chemistry and the number of polymer chemistry departments in the Japanese universities. For note on method of computation use Appendix A. Source: Yamada (1982).

1 Research Papers Published in Foreign Country
2 Research Funds
3 Research Papers
4 Patents

5 Production
6 Researchers
7 Number of Polymer Chemistry Departments in the Japanese Universities

by establishing their own firms. (The same sort of thing occurred much later in the case of the scientific instrument industry.) Sometimes these chemist-inventors worked with partners who had greater commercial experience, sometimes they acquired this knowledge themselves, but they laid the foundations for the modern chemical industry.

As, however, the new chemical firms grew up and consolidated their new markets, some of them (especially the German dyestuffs firms) made an extraordinarily important social innovation — the captive industrial R and D laboratory. This meant that they were no longer so vulnerable to the 'creative destruction' brought about by exogenous science and technology through new entrepreneurs. They themselves learnt the trick of institutionalizing this process. By earning exceptional profits on their major innovations they were able to finance scientific and technical activities on such a scale as to retain the ability to generate successive new waves of invention and innovation, or at least to keep fairly close behind the leaders (Figure 2.4).

Nylon accounted for more than half the corporate profits of Du Pont for many years, and such products as polyethylene and PVC made a similar contribution to ICI and the German chemical firms. The original growth and strength of the German chemical industry was based on the profits from synthetic dyestuffs; its later growth on those from drugs and plastics. The evidence of our studies on dyestuffs, plastics and organic intermediates confirms the view of those who have argued that in these sectors the largest chemical and oil firms have dominated the inventive and innovative process over the past fifty years whilst strengthening their own market leadership. Not only did the German chemical firms BASF, Bayer and Hoechst rise to their pre-eminent position in the world through their dyestuffs inventions in the nineteenth century, but it was also I. G. Farben and ICI who were responsible for most of the major dyestuffs inventions since the 1920s.

In the case of plastics and synthetic fibres it was the leading chemical firms in each country (I. G. Farben in Germany, Du Pont in USA, ICI in Britain, and Montecatini in Italy) who were able to dominate the technical developments leading to the introduction of most of the major new materials and their application in a great variety of novel end uses. They were also usually the first to imitate or license each others' major innovations (nylon 66, nylon 6, polyethylene, PVC, acrylics etc., see table 5.1). The extraordinary 'bunching' of I. G. Farben inventions and innovations in the period from 1925 to 1945 is particularly notable.

Table 5.1 Patents and innovations in synthetic materials (percentage world total)

Patents and innovations	Total	Percentage world total		
		I. G. Farben*	Du Pont	ICI
All plastics patents taken out by firms 1791–1945	6777	20	6	2
All plastics patents taken out by firms 1931–45	4341	20	8	2
'Major technical advances' in patent literature 1791–1945	117	26	10	6
Innovations in synthetic materials 1870–1945	56	32	9	2
'Major innovations' 1870–1945	20	45	10	5
Innovations 1925–45	36	44	11	3
First 'imitations' 1870–1945		14	4	8

*Including predecessors and successors.
Source: Freeman (1974), p. 93.

We have suggested two main mechanisms to explain this bunching. One of these was the advance in fundamental science in the 1920s associated with the work of Staudinger on the structure of long-chain molecules and the establishment of a recognizable 'polymer science' during that period. The 'demand' for synthetic rubbers, fibres and other materials was already strong during World War I but it could not be met then because of the lack of this essential pre-condition. Once this condition was satisfied, however, then the other mechanism could exert its strong influence. The autarchic pressures from the demand side associated with German re-armament and the German war economy affected the synthetic rubbers especially.

5.4 The growth cycle of a new technological system

Finally, we shall attempt to relate this discussion of science, technology and demand in the plastics industry to the cyclical development of the industry more generally, and specifically to the issue of employment:

(i) On the basis of the two mechanisms described above (the establishment of polymer science and autarchic pressures) there was a take-off in German plastics innovation and production in the late 1920s and 1930s. This was followed by other leading industrial countries (USA, UK, Japan etc.) with a lag of a few years but the world-wide diffusion process took place mainly in the post-war period.

(ii) Re-armament and war provided a forcing house for technical development and the rapid scaling up of production, especially for synthetic rubber in both Germany and USA, but also for fibres and other materials, especially polyethylene (radar and cables) and PVC (many applications in Germany). Many new basic innovations took place both in products and processes and there was also a surge of secondary invention and discovery. Recorded patent applications declined temporarily in the 1940s, however, because of war-time security practices and other effects of the war (Figures 5.1 and 5.2).

(iii) There was a temporary slow down in the tempestuous growth of the industry immediately after the war in Germany and Japan, where bomb damage, occupation measures and the dissolution of the I. G. Farben trust caused delay. Especially in the USA, however, new civil applications were very rapidly developed and the 'swarming' process set in strongly in the 1950s.

(iv) Arrangements and understandings between some of the leading chemical firms had led to some restrictions on the licensing and transfer of the new technology. For example, ICI had licensed Du Pont to produce polyethylene during the war and these firms had also gained some access to German technology during the occupation and reorganisation of I. G. Farben. However, the high profits being made from polyethylene by Du Pont and ICI led to considerable pressures from would-be swarming imitators to acquire licenses. After a long legal battle the Supreme Court finally required ICI and Du Pont to license a number of other American chemical and oil firms.

(v) The swarming process now gathered force for this and other synthetic materials. New plants were being built all over Europe, USA and Japan and their size increased rapidly. The shift from coal to oil as the main feedstock for the industry — in which the USA led the way — greatly facilitated the introduction of flow processes, the scaling up of plant and further basic process innovations. Moreover, the use of oil as feedstock meant that 'building block' chemicals such as ethylene and styrene could now be easily supplied in large quantities from oil refineries and adjacent petro-chemical plants. In addition to the attraction of the very high profits being made by the innovators, the oil companies now had an additional motive for entering the plastics industry and many of them did so. Because of the large amounts of capital required to build new capacity, they and the large chemical companies constituted most of the 'swarm'. This also meant that total numbers of producers remained relatively small. However, their entry was sufficient to erode the profit margins of the industry. Some of the oil companies made hardly any profits from plastics, even in the 1950s and 1960s, but stayed in the business because of their desire for long-term diversification. The chemical companies began to complain of their falling margins.

(vi) With the swarming process and the continuing intensive R and D by the original innovators, there was an immense surge of patenting and publishing activity (Figure 5.2). As we have seen these were closely inter-related and both stimulated and followed the innumerable new applications that were opened up by the applications research, marketing activities and technical service departments of the producers. Strong 'virtuous circle' mechanisms were now in operation with demand and technology mutually stimulating each other. Growth rates were commonly between 15 and 20 per cent per annum for output in the leading industrial countries, and even in countries such as Sweden, with little or no production, consumption increased at a similar rate. The machinery industries also contributed to the speed of technical advance but the large chemical companies continued to account for the greater part of it. A rather more significant contribution came from the plastics fabricating industry, in which the swarming of small firms (Schumpeter mark I) was much more important, together with new departments of large engineering firms and other big users of plastic components.

(vii) The new technology gave rise to the growth not just of the synthetic materials industry, but a fabricating industry which employed about three times as many people, and a smaller machine-building industry. As the technology became more standardized and more widely diffused, the design and construction of new plants was taken over increasingly by a new group of firms — the chemical engineering contractors. Originally this industry came into existence mainly to serve the oil companies in the design and construction of refineries and related facilities, but they now increasingly moved into organic chemicals, fertilizers and plastics. They specialised in the provision of 'turn-key' packaged plants world-wide and insofar as they conducted new R and D work, this was mainly related to scaling up and modification of the designs that they acquired under license from the large chemical companies. Sometimes they were actually subsidiaries of the large chemical firms (e.g., Uhde is a subsidiary of Hoechst), but more commonly they were independent firms. The emergence of this industry and its role in new plant construction was one indication of the world-wide tendency to standardization of a mature technology. A few of the exceptionally innovative contracting companies did make some original basic process innovations as well as many improvements, but the share of the large chemical companies in patents and innovations continued to be dominant (Freeman *et al.* 1968). The emphasis of R and D however shifted increasingly to process technology rather than the materials themselves and this was reflected both in the numbers of scientific papers and of patents (see figure 5.5).

(viii) Finally, however, the 'retardation' factors which have been discussed in Chapter 4 came increasingly into play during the 1970s. Surplus capacity began to emerge on a serious scale, especially in synthetic fibres, and a more general situation of low profitability became widespread. Growth rates slowed down markedly and the indicators of patenting and publication also decelerated or declined (Figure 5.2). Among the retardation factors affecting the growth cycle of the industry, the following were probably the most important:

(a) Market saturation. During the 1970s the total volume of plastics production was greater than that of all non-ferrous metals combined and equal to a significant fraction of total steel production. Synthetic rubber accounted for about three quarters of the combined consumption of natural and synthetic rubbers. Synthetic fibres accounted for about half of all fibres consumption. Whilst the new properties of synthetic materials did stimulate many completely new applications, a large part of their growth was at the expense of older natural materials (paper, cotton, rubber, metals, jute, etc.) which had much more sluggish growth rates. It was inevitable that after an initial period of very rapid market penetration based on technical and price competition, the growth rate of synthetics should asymptote towards the over-all growth rate of materials.

(b) Tendency of input prices to increase with the expansion of the industry. Labour costs in many countries were rising rapidly in the 1960s but in this industry it was the cost of capital, of energy, of new plant and of feedstock which were probably the most important elements in the squeeze on profit margins

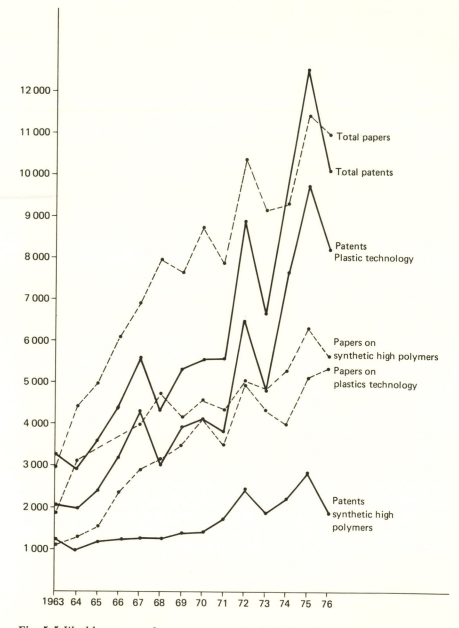

Fig. 5.5 World papers and patents on synthetic high polymers and plastics technology. Source: Walsh *et al.* (1979).

apparent already in the late 1960s, but more obviously since 1973. In the fabricating industry labour costs were much more important.

(c) Slow down of technical advance. Although we have stressed that basic innovations continued to occur in the 1950s and 1960s, and that the rate of secondary and induced invention continued at a very high rate, there were indications of diminishing returns to further investment in R and D and other technical activities and of a marked shift to process inventions. Scaling-up of plant appeared to reach some technical and economic limits in the 1970s and this had been one of the main sources of productivity advance, because of the operation of Chilton's Law. The limits of new applications research also began to be apparent. Technical advance has certainly not stopped and the possibility of further basic innovations remains, but some retardation is evident from most of the available indicators.

(ix) So far as employment is concerned the new industry generated a rapid growth of employment, especially on the fabrication side from the 1930s to the 1960s. Even though production of the materials themselves was capital intensive from an early stage, the high rate of introduction of new plants and the supporting scientific, technical and marketing services gave rise to many new jobs, making this (together with drugs) the main source of new employment in the chemical industry. Over 1 million people were employed in the fabricating industry, the materials industry and the associated machine-building industries in the USA, and rather more in the EEC by the late 1960s. However, because of technical advance associated with economies of scale (Chilton's Law) the employment generated for each new wave of investment tended to decline. Labour productivity grew very rapidly in the 1950s and 1960s, and by the 1970s the effects of the retardation factors which have been discussed led to stagnation or even decline in employment.

(x) The growth cycle of this new technological system appears to exemplify fairly well the concept of a half-century cycle affecting the economies of the OECD countries more generally, although of course with national variations. How far in combination with other industries it was the cause of such a cycle, or simply the result of it, and how typical it was, will be discussed in later chapters. This does not mean, of course, that no further growth of plastics production will occur. There are still possibilities of considerable further growth and of further inventions and innovations.

But the exceptional impetus that the industrialized economies derived from this new technological system in the past half century can no longer be sustained.

6 ELECTRONICS

We have stressed the importance of clusters of inter-related inventions and innovations in the growth cycles of new industries. We have called these inter-related clusters and their diffusion through the economy 'new technological systems'. The phenomenon in which we are interested is something more than the life cycle of a particular industry, although it certainly includes this. Nor is it just the emergence of a new scientific discipline or sub-discipline, although it may partly derive from this. The concept implies an embodied technology and not just a set of new scientific ideas. We have used this expression to capture the fact that major new technologies affect many industries to varying degrees and we are interested especially in those that have the widest effects on the economy — the 'general natural trajectories' of Nelson and Winter, and Kierstead's innovations of 'wide adaptability'. The economic system cannot digest a major new technology overnight or over a decade — the responses required are too many, too varied and often have very long time lags.

In this chapter we shall be discussing a new technology system with even more pervasive effects throughout the economy than synthetic materials — electronics. From the outset we can see that this is a very different case from synthetic materials and some people would maintain that it has now reached its period of most rapid development and shows no sign of maturity after eighty years of sustained growth of output at a rate often exceeding 15 per cent per annum. Others would say that it exhibits elements of all four of van Duijn's 'variations' in cyclical behaviour (as illustrated in Figure 6.1) and this is true so far as particular products or sub-sectors are concerned (e.g., colour television substituted for monochrome television during the 1960s rather on the lines of Figure 6.1(c)).

Nevertheless, we shall endeavour to show that several key sectors of the electronics industry have experienced a post-war growth cycle not altogether dissimilar from the general pattern we have described. In particular, as the industry has become more capital intensive, the rate of growth of employment has diminished and the absolute level of employment has actually declined or stagnated in some sectors even where sales and investment continue to expand.

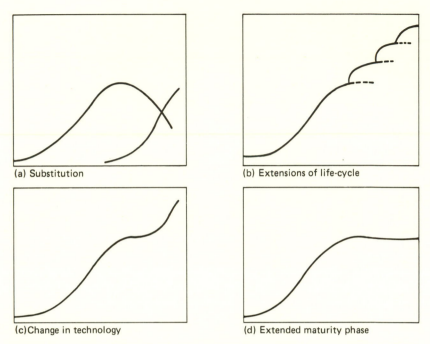

(a) Substitution   (b) Extensions of life-cycle

(c) Change in technology   (d) Extended maturity phase

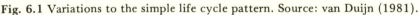

Fig. 6.1 Variations to the simple life cycle pattern. Source: van Duijn (1981).

At the same time there are new growth points for example in micro-computers, where employment is growing very rapidly.

6.1 The electronics industry

As it is usually defined, the electronics industry 'proper' involves the production of components and their assembly in a variety of end products including consumer goods (e.g. radio, television and tape recorders) and capital goods (e.g., radar, computers or scientific instruments and military equipment). However, in practice it has proved difficult to follow any classification consistently and no country has succeeded in keeping statistics that follow close behind the extremely rapid rate of technical change. One problem is the re-classification of products, formerly classified to other industries, which now embody electronic circuits replacing to varying degrees electro-mechanical systems. Major examples are computers, calcula-tors and telephone equipment, which are now usually (but not always) classified with electronics. Many products include mechanical, electrical and electronic components or subsystems. Machine tools were once entirely mechanical, later became largely electro-mechanical and now increasingly incorporate electronic control systems. Radio and television, which have always been classified with electronics, have of course always included the assembly of many purely mechanical or electrical components — and it is partly an arbitrary decision or a matter of convenience as to which products are reclassified, as electronic systems increasingly penetrate a wider range of end-products. Electronic component firms are now major manufacturers of watches and toys and the scientific instrument industry is now largely an electronic one. More than half the cost of some military aircraft is in the electronics. The important point is that the influence of electronic technology in the economic system is more widespread than might appear simply from some of the available statistics, and this is one of the main reasons for our use of the concept of 'new technological systems'. The industry thus narrowly defined now employs well over 1 million people in the USA and similar numbers in the EEC; much greater numbers of jobs were generated in the post-war period in service applications, repair and maintenance work, computer installations and bureaux and so forth.

The critical components in electronic circuits are the so-called 'active' components — valves and transistors — which direct, control, amplify or otherwise influence the flow of electricity through the circuit. It was technical innovation in valves in the early part of this century, and in transistors and solid state technology in the second half of the century, which formed the basis for the remarkable rate of technical change throughout the industry. Successive 'generations' of components gave rise to a rapid succession of redesigned electronic products in some sectors rather on the lines of Figure 6.1(b). The so-called passive components continue to account for a high proportion of total components output and technical change has been important here too, but the key innovations have been in the active components and in the integration of entire circuits. In this chapter we shall first of all discuss the 'older' parts of the industry — radio and television — then the electronic computer and finally the revolutionary developments in solid state technology, culminating in micro-electronics.

6.2 Radio and television

The first branch of the industry was radio and in the period before World War II the two were almost synonymous. The radio industry took off about 1901 with Marconi's early demonstrations of the feasibility of transmitting signals over long distances and from ship

to shore. His achievements as an innovative technologist, who sometimes defied received scientific opinions, should on no account be
under-rated. Nevertheless, radio was based on fundamental advances
in physics in the latter part of the nineteenth century (Maxwell,
Hertz, Branly, Popov) and a cluster of basic inventions at that
time — including both radio itself and several key components,
especially the diode valve (invented by Fleming 1904) and the
triode valve (De Forest 1907). Sturmey (1958) maintains that
commercial demand played no part in these early valve inventions.
Early applications of radio were in long-distance communications
and had the effect of drastically reducing cable rates and generally
improving communication with ships. However, it was with broadcasting in the 1920s that large-scale swarming began, with many
new entrants to the industry followed by the typical Schumpeterian
pattern of declining profit margins, bankruptcies and concentration
in the 1930s. The strongest surviving firms, such as RCA, Philips,
Telefunken, Marconi and EMI were characterized by strong R and D,
especially in valve technology.

Thus the radio industry followed a long-term cycle of growth
and maturity extending over about forty years, but the basic innovations on which it was built up were introduced in the 'prosperity'
phase of the third Kondratiev and the main swarming took place
late in the cycle. However, a second and more important cycle of
growth for electronics began with several revolutionary product
innovations in the 1930s and 1940s (especially television, radar
and the electronic computer) and a whole series of revolutionary
innovations in active components and circuits starting with the
transistor which also gave a renewed impetus to radio itself.

The growth cycles of the television and radar industries fit fairly
well into the fourth Kondratiev long-wave pattern of very rapid growth
in the 1940s and 1950s, moving into a mature phase of high capital
intensity, market saturation and stagnant or declining employment in
the 1970s. Radar was always deeply affected by military demand and
for this reason it took off much more rapidly during World War II
and the Cold War, and is still affected strongly by the vagaries of this
demand — whereas the growth of television was delayed until civil
broadcasting could be resumed after the war. Once this occurred
(with the licensing of the original technology of RCA, EMI and
Telefunken) the saturation of domestic markets in the leading industrial countries took little more than a decade and, as with radio,
the transition to a replacement market led rapidly to the erosion of
profit margins and a concentration process. The number of producers
was never large and it was falling in the 1960s and 1970s.

The pattern of the industry's development corresponded to Schumpeter's mark II model since both the original innovators and the imitators were large established producers of radio and other electrical goods with strong R and D facilities. Colour television (in which RCA and Telefunken were again the leading innovators) followed a similar product cycle in the 1960s–1970s (see Figure 6.2), but by this time the international diffusion of the technology meant that the swarming process was dominated by Japanese firms whose competition proved devastating for the older American and European industries. Japanese firms were not among the original innovators but by 1977 Japan accounted for over half of world production in colour television and three quarters of world exports. They were exporting about 5 million sets, compared with about 1 million from Germany and 250 000 from the UK, despite the fact that Japan was still limited in many markets in Europe by the PAL patents and other restrictions. Later in the 1970s Japanese direct exports declined, especially to the USA, because of Japanese investment overseas and because of agreements between Japanese and American producers.

As in the case of the introduction of transistors into the radio industry, this extraordinary Japanese success was not based on simple carbon copy imitation, but involved a whole series of product improvements and process innovations. After comparing the performance of the American, European and Japanese industries, Sciberras (1980) concluded:

Japanese firms have been the most successful innovators [in the 1970s]. By applying advanced automation in assembly, testing and handling to large production volumes, the Japanese have achieved drastically superior performance in terms both of productivity and of quality.

Although he found that the main advantage of the new automated techniques was in improving product reliability, Sciberras calculated that Japanese man-hours per set were 1.9 compared with 3.9 in Germany and 6.1 in the UK. He attributed the opening up of this remarkable productivity gap mainly to the integrated approach to automation technology and to the intensive training of personnel at all levels in Japanese firms. Peck and Wilson (1982) also point out that the Japanese manufacturers were the first to introduce integrated circuit technology into the colour television industry (with the important economies in assembly labour that this involved). The success of this innovation was based on a joint research effort starting in 1966 and involving five television manufacturers, seven semi-conductor manufacturers, four universities and two

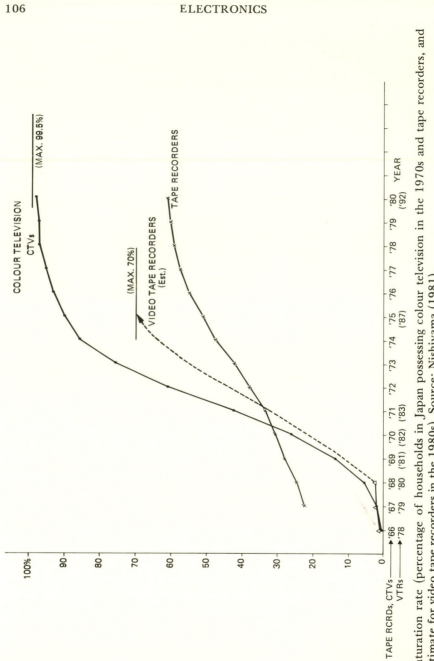

Fig. 6.2 Saturation rate (percentage of households in Japan possessing colour television in the 1970s and tape recorders, and estimate for video tape recorders in the 1980s). Source: Nishiyama (1981).

research institutes and the overall backing of MITI (the ministry for trade and industry). This example illustrates the capacity of the Japanese social system to achieve a flexible mobilization of resources to make and diffuse decisive innovations quickly.

We shall return later to this Japanese success in discussing the changing international leadership in successive waves of new technology in Chapter 9, but for the moment we note that their extremely effective world-wide competition led to a more rapid decline of profits and employment in some of the older producing countries in the 1970s. Radio and television were already mature industries in the 1970s and would in any case have experienced slower growth or even some decline in their labour force. The product cycles of television and colour television fit fairly well into the general postwar long wave, but product innovations such as video tape recorders (Figure 6.2), Prestel sets etc. may permit a pattern of development for parts of the industry on the lines of curves (b) or (d) in Figure 6.1. The existing capacity and employment of the industry appears, however, quite capable of handling this demand.

6.3 The computer

It was the innovation of the electronic computer, in combination with a whole cluster of innovations associated with solid state technology, which vastly extended the range of applications of electronics in control systems, information systems and telecommunication systems. The convergence of these associated technologies was the most important development of the fourth Kondratiev and may provide one of the elements for a fifth expansionary wave. The 'sun' at the centre of this whole constellation of associated inventions and innovations was the electronic computer.

Attempts to build mechanical or electro-mechanical computers had been made ever since Babbage's prototype in the 1830s and these efforts greatly increased during World War II. Both radar and code cracking made use of early computing devices. It was the German inventor, Zuse, who (after he had already designed several electro-mechanical machines at the Charlottenburg Technical High School) first succeeded in designing a fully fledged electronic model in 1943. The much larger ENIAC computer, designed and constructed at the University of Pennsylvania under a contract from the Army Ballistics Laboratory, began to work in 1945. Several other machines were designed and operated at various British and American universities and government laboratories in the late 1940s. In these early days the work was sustained by the enthusiasm of the groups of scientists and engineers who were most closely involved and the long-term sponsorship of a variety of government agencies.

Katz and Phillips (1982) have recently published a thorough account of the early history of the US computer industry and make particularly interesting comments on the reasons why private funds were not committed to the commercialization of the electronic computer:

> . . . the general view prior to 1950 was that there was no commercial demand for computers. Thomas J. Watson Senior, with experience dating from at least 1928, was perhaps as acquainted with both business needs and the capabilities of advanced computation devices as any business leader. He felt that the one SSEC machine which was on display at IBM's New York offices 'could solve all the scientific problems in the world involving scientific calculations'. He saw no commercial possibilities. This view, moreover, persisted even though some private firms that were potential users of computers — the major life insurance companies, telecommunications providers, aircraft manufacturers and others were reasonably informed about the emerging technology. A broad business need was not apparent.

Again, therefore, we have to qualify the theory of demand-led invention with important reservations about the importance of basic research and technology push in relation to revolutionary innovations. At the same time, as in the case of plastics, we should note the role of military interest in promoting developments during and after World War II. Katz and Phillips sum up this peculiar confluence of scientific interest and potential applications:

> In a real sense the technologist users (in government) and the technologist suppliers (in private firms) had coincident interests and were members of a cognisable 'fraternity'. They attempted to prevail on their respective host organisations to supply funds to meet their technological and scientific objectives. The demand, that is, was more in the form of budget requests by this group for funds for investment in R and D — without regard to immediate economic returns on investment — than it was a demand for marketable computer hardware (p. 15).

At this time, Eckert and Mauchly were among the few believers in a wider commercial market and they made intensive efforts to interest private companies and financiers after their dismissal from the University of Pennsylvania in 1946 because of their financial interests. They met with very great difficulties and disappointments and were destined for bankruptcy when Remington Rand acquired their corporation in 1950. However, after this acquisition the prospects were still thought to be so dubious that Remington Rand attempted to cancel all outstanding contracts for their computers.

Only a few government organisations refused to cancel, such as the Bureau of the Census.

Whilst these great difficulties were being experienced in the early commercialization of the electronic computer, enormous progress was being made with technical improvements, resulting in many clusters of basic and secondary inventions affecting especially the development of stored programmes, memory devices, and programming languages. The early days of the technology were characterized by very free interchange of technical and scientific information, including many joint seminars involving most of the interested scientists and representatives of government and industry. 'Software' and some other inventions could not be patented, but a surge of patents and publications began both in the USA and the UK which has continued to this day. Again, this upsurge preceded the growth of the commercial market and in Britain the NRDC organised a patent pool.

The Korean War further stimulated US government interest and led to IBM's full involvement. Although IBM was by no means the first innovator, once the company had to meet contracts for a small number of machines for military orders it began to take a serious interest in entering the civil market. Even so, when the Applied Science group at IBM later proposed launching the 650 computer forecasting a sale of about 200 machines, the Product Planning and Sales Department forecast no sales at all. Thomas Watson Junior took the side of the scientists and in the end 1800 machines were sold, mainly for business applications.

Katz and Phillips describe the 650 as the 'Model T' of the computer industry and in many senses it was, although the transistorized 1401 launched in 1958 might have stronger claims. It marked the first big exploitation of economies of scale, and it was relatively cheap and easy to maintain. It also demonstrated to IBM (and its competitors) that the industry could be extremely profitable. IBM was already a highly successful company in the field of tabulators and other office equipment and it was probably this unrivalled knowledge of the market and strong sales and service organization that gave the company the edge over all its competitors. However, it also had great technical strength and a very strong commitment to both R and D and education. The combination of these characteristics enabled the company to become, for a long period, one of the most profitable and fastest growing companies in the world. Gross income grew from $116 million in 1946 to $8273 million in 1971. Estimates vary slightly, but most of them agree that in the 1950s, 1960s and 1970s IBM accounted for about 60–70 per cent of the value of all computer sales throughout the world. In terms of numbers of computers installed the IBM share would be a little less, as other smaller companies (such as DEC), were more successful with mini-computers. IBM employment world-wide grew from 22 000 in 1946 to about 165 000 in 1965 and 300 000 in 1974; it has since tended to level off (the 1980 figure was 340 000). Although 'swarming' did take place in the 1950s and 1960s and a number of large electronic companies, such as RCA, General Electric, AEI, Siemens, English Electric, EMI, and Raytheon, entered the industry, few of them had much success and none made anything like the profits of IBM. Only in 'niche' areas, such as special applications and mini-computers, was IBM's predominance seriously challenged. Efforts to sustain competitive national firms in several European countries (especially Britain, France and Germany) also met with great difficulties; many firms including some of those mentioned above fell off the bandwagon and withdrew from the industry in the 1960s, having failed to make any profits both in the USA and in Europe. The computer industry over this period was thus an extreme example of Schumpeter's mark II model. The situation has now changed as a result of the developments in the semiconductor industry and the innovation of the micro-computer and micro-processor in the 1970s. A new wave of small firms has now entered the industry such as Sinclair with the ZX81; these developments are further discussed in Chapter 7.

However, IBM is already one of the world's biggest chip manufacturers and whether its predominance can be seriously challenged remains to be seen. We now turn to developments in the component industry.

6.4 Solid state technology

The need to have large numbers of valves in some types of electronic equipment was already creating serious problems in the 1930s and their miniaturization was faced with inherent physical limitations. Large numbers of valves involved very high power consumption and generated a great deal of heat, as well as occupying an enormous space. The problem of heat in turn led to problems of reliability. Hence the first generation of computers in the 1950s, although successful (even with valve technology) appear today to be extremely cumbersome, expensive and environmentally awkward. In fact, the limitations of valve technology were realised from the start, and Lukoff (1979) noted that 'In 1953 Univac realised that the days of the vacuum tube were numbered. About 90 per cent of all computer maintenance problems were due to the vacuum tube and it simply had to be replaced'.

It was also these well-known limitations of valve technology in relation to telecommunications that led Bell Laboratories to sponsor a research programme in solid state physics in the 1940s, which ultimately led to the transistor. This programme was able to draw on much earlier academic research as well as on wartime experience with crystal diodes and high purity germanium and silicon. A solid state division of the American Physical Society had been set up by 1947, and the links between transistor technology and fundamental scientific advances in physics have remained extremely strong. This has been shown particularly clearly in a study entitled *Science–Technology Coupling in Electronics* by Lieberman (1978). He found that for industry authors publishing in the *Transactions of the Institute of Electrical and Electronic Engineers*, references to the basic science literature from *Physical Review* were actually slightly more recent than those of university authors in basic science. He concluded that the science–technology coupling had remained strong in electronics over more than two decades as a result of the continued birth of new science-related technologies in the solid state field. Bell Laboratories was in an almost unique position to make these strong connections with basic physics research, because although an industrial laboratory, it conducted fundamental research programmes on a long-term basis and operated almost like a university for industry.

The need for smaller, more compact and reliable active components was a general one and was not confined (as is sometimes believed) to military or space applications. Indeed the success of the first transistor was concealed from the military until shortly before publication in order to prevent them from trying to put a security blanket over the new technology. Possibly because of its position as the research arm of a major government-regulated utility (American Telephone and Telegraph), Bell followed a very open policy in communicating its results and licensing all comers and, following the first innovation, progress in miniaturization has been extraordinarily rapid (Figure 6.3). A decisive stage was the invention of the integrated circuit in the late 1950s, enabling a number of active and passive components to be produced together on one 'chip'. Although G. W. Dummer of the Royal Radar Establishment was one of the first to envisage this possibility and a model was demonstrated at Malvern in 1957, the first patent was taken out by J. Kilby of Texas Instruments in 1959. The first commercial

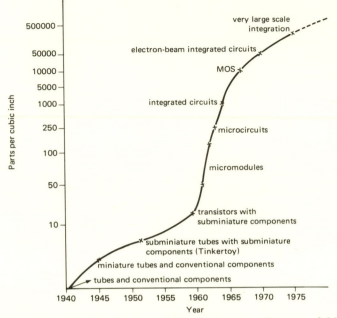

Fig. 6.3 Electronics miniaturization 1940–75. Source: Braun and MacDonald (1978).

innovation based on the much more effective planar production process was launched by Fairchild in 1962. The average price of integrated circuits fell from $50 in 1962 to $1 in 1971 and the displacement of valves as active components during this period is shown in Figure 6.4. Since then a series of further product and process innovations (Tables 6.1, 6.2 and 6.3) have made it possible to increase the number of components on a chip so rapidly that it is now approaching a million (VLSI: 'very large scale integration).

By the early 1970s this meant that it was possible to assemble all the components needed for a powerful central processing unit of a computer on a single chip — the microprocessor. The 'hardware' costs of computing fell by more than two orders of magnitude. The marginal cost of additional chips is so low that additional standard units are available at a fraction of the cost of the older computers and can be distributed by ingenious design throughout automation and control systems for all types of industrial and office process. These remarkable technical advances which in combination permitted great reductions in costs and improvements in technical performance, were achieved partly by the large well-established firms in the electronics industry and partly by new

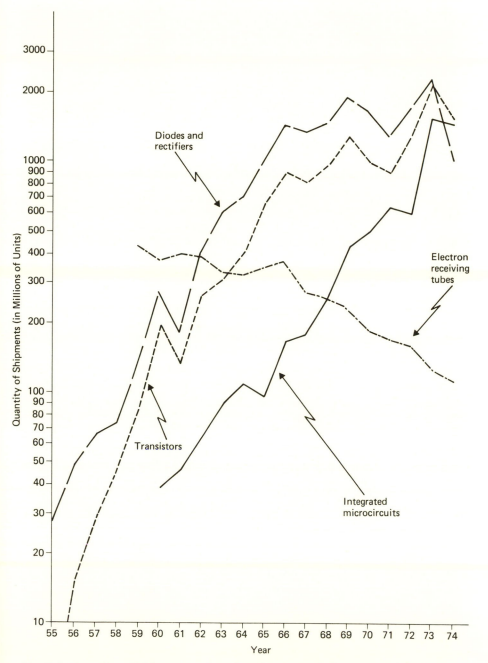

Fig. 6.4 Quantity of shipments of semiconductors and electron receiving tubes (1955–74). Source: US Federal Trade Commission (1977).

Table 6.1 First commercial production of various electronic components

Component	USA		UK		Italy		Western Germany	
	Date	Firm	Date	Firm	Date	Firm	Date	Firm
Silicon mixer diode (point contact)	1938	GE	1939	BTH*			1943	Telefunken*
Germanium diode	1941	GE*	1942	BTH			1943	Telefunken*
Point-contact germanium transistors	1951–2	Bell*, WE	1953	AEI, STC*			1954	Telefunken*
Silicon junction diode	1952	Bell,* WE	1954	Ferranti	1959	SGS		
Germanium junction transistor	1951	GE, RCA	1953	Mullard†	1954	Philips	1953	Valvo / Telefunken
Silicon-grown junction transistor	1954	TI*	1957	TI†	1958	Philips	1959	Valvo
Alloy-diffused germanium transistor	1955	Bell, WE	1958	Mullard†	1960	Philips	1960	AEG*
Silicon controlled rectifier	1956	GE*	1957	AEI*	1960	SGS	1957	Siemens*
Silicon power rectifier	1955	GE*	1957	AEI,* STC			1958	AEG*
Germanium power rectifier			1954	AEI				Siemens
Diffused (mesa) silicon transistor	1958	TI*	1959	TI, Ferranti	1959	SGS		Siemens
Diffused (mesa) germanium transistor	1956	TI	1958	TI				
Planar transistors	1960	Fairchild*	1961	Ferranti	1962	SGS†	1963	Valvo
Epitaxial transistor	1960	WE	1962	TI, STC			1963	Intermetall / Valvo
Integrated monolithic microcircuits	1961	TI*	1963	TI† Ferranti*	1966	SGS†	1964	Telefunken

*Own R and D.

†Under licence.

SGS = Societa Generale Semiconduttore (partly owned by Olivetti, Teletra and Fairchild); GE = General Electric; WE = Western Electric (WE forms the manufacturing function and Bell Labs the research function of AT & T = American Telephone and Telegraph); RCA = Radio Corporation of America; TI = Texas Instruments; BTH = British Thomson Houston (later part of AEI and of GEC/AEI); STC = Standard Telephone Company (British subsidiary of ITT); Valvo and Mullard are subsidiaries of Philips.
Source: Freeman (1974).

Table 6.2 Major product innovations in the semiconductor industry since the integrated circuit

Innovations	Firm	First commercial production
MOS transistor	Fairchild	1962
DTL integrated circuit	Signetics	1962
Gunn diodes	IBM	1963
Light-emitting diodes	Texas Instruments	1964
TTL integrated circuit	TRW	1964
MOS integrated circuit	General Microelectronics	1965
	General Instruments	
Magnetic bubble memory	Western Electric	
MOSFET (MOS field-effect transistor)	Western Electric	1968
	Philips	
Schottky TTL	Texas Instruments	1969
CCD (charge coupled device)	Fairchild	1969
Complementary MOS	RCA	1969
Static RAM	Intel	1969
Silicon-on-sapphire (SOS)	RCA	1970
P-MOS		1971
3-transistor cell dynamic RAM (1K bits)	Intel	1971
CMOS		1971
Microprocessor	Intel	1972
I^2L integrated circuit	Philips	1973
1-Translator cell dynamic RAM (4K bits)	Intel	1974
VMOS integrated circuit	AMI	1975
C^2L integrated circuit		1976
MNOS		1976
Micro-computer (8048)	Intel	1977
V-MOS	Mitsubishi	1978
64-K bits memory	Fujitsu	1978

Source: Dosi (1981).

entrants to the industry in the USA (Tables 6.1, 6.2 and 6.3). The semi-conductor industry is often quoted as one which supposedly demonstrates the overwhelming importance of new small firms overcoming the managerial and technical conservatism of large established firms with their commitment to older products and processes. Thus it might be construed as a throw-back to a 'Schumpeter mark I' pattern of growth in the 1950s. Certainly new young firms, such as Texas Instruments, Fairchild (later the 'Fair Children') and Mostek and Intel, did make an extremely important contribution to the advance of the technology, especially in innovating new processes and scaling them up to permit huge price reductions. However, this must be qualified on three grounds.

First, an examination of Tables 6.1 and 6.2 shows that that large electronics firms with very strong R and D were responsible for about half the major innovations from 1950 to 1980 (General Electric,

Table 6.3 Major process innovations in semiconductor industry

Innovation	Firm	Date of Development
Single crystal growing	Western Electric	1950
Zone refining	Western Electric	1950
Alloy process	General Electric	1952
3–5 Compounds	Siemens	1952
Jet etching	Philco	1953
Oxide masking and diffusion	Western Electric	1955
Planar process	Fairchild	1960
Epitaxial process	Western Electric	1960
Plastic encapsulation	General Electric	1963
Beam lead	Western Electric	1964
Dielectric isolation	Motorola	1965
Collector diffusion insolation	Western Electric	1969
Ion implantation	Mostek	1970
Self-aligned silicon gate	Intel	1972
Integrated injection logic	Philips	1973
Vertically oriented transistor	AMI	1975
Double polysilicon process	Mostek	1976
E-beam mask projection		1976
Plasma nitride processing		1976
Automatic bonding on 'exotic' (35 mm film) substrate	Sharp (Japan)	1977
Vertical injection logic	Mitsubishi	1978

Source: Dosi (1981).

RCA, Phillips, and of course Bell with its manufacturing arm, Western Electric). They were responsible for more than half of all the patented inventions (Dosi 1981) and process innovations (Table 6.3). Table 6.1 shows that in Europe (as compared with the USA) both the early imitators and first innovators were almost entirely the large electronics firms. Table 6.2 covering the innovations since the integrated circuit does not show first imitators in other countries, but the same point is still valid.

Secondly, some peculiar features of the swarming process should be taken into account. Many of the original 'spin-off' companies which have been such a feature of this industry were formed by individuals or groups of scientists and engineers who left Bell Laboratories and other large companies (Figure 6.5). As a regulated telephone utility the Bell (ATT) system was not entirely free to develop in a normal way as a component manufacturer and, as we have seen, followed a policy of giving very free access to technical information to other firms in the electronics industry.

Thirdly, there was quite a marked difference between the pattern of swarming in the USA and that in Europe and Japan. In Europe very few new small firms or spin-off firms entered the

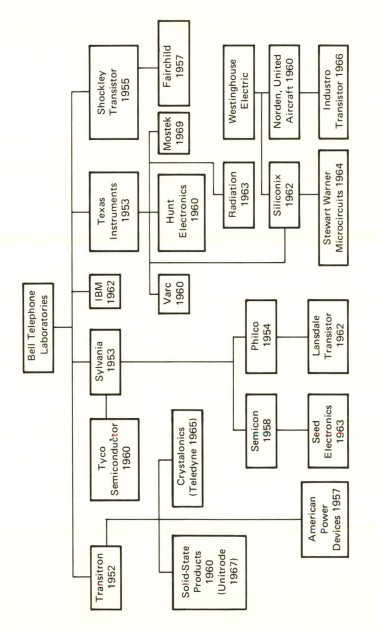
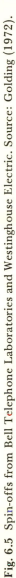

Fig. 6.5 Spin-offs from Bell Telephone Laboratories and Westinghouse Electric. Source: Golding (1972).

semiconductor industry, so that the international diffusion of the new technology was handled almost entirely either by subsidiaries of firms in the USA or by the large established electronic companies (such as Siemens, AEI, GEC, CSF, Phillips and its subsidiaries, etc.). Even in this industry Schumpeter's mark II model is relevant. New entrants such as Texas Instruments and Fairchild were able to make large profits from their innovations, but for rather short periods by rapidly increasing their market share through a combination of innovations, economies of scale and price reductions. Perhaps the biggest single difference between the USA and European industries was the way in which large orders for new circuits from the military and space agencies enabled both new and older firms to exploit economies of scale and the associated process innovations (Golding 1972). Whereas innovators in other industries (such as drugs) have often initially attempted to follow a high price and high profit strategy during the period of patent protection, the semiconductor industry was remarkable for its price falls. Nevertheless, Sciberras (1977) maintains in his study of the pricing behaviour of the most dynamic new firms that they were following a strategy of profit maximization. In his view the motive for big price reductions was to deter new potential entrants by creating scale barriers of entry in addition to technological entry barriers. In this, they have been partly successful but the 'competing away' of profits has nevertheless pushed the average rate of profit down to the general level for manufacturing in the USA. Dosi (1981) concludes that the world semiconductor industry has in fact now become a mature international oligopoly.

Employment in some mature sections of the electronics industry, such as television, has levelled off or declined and although growth continues in some of the most dynamic sectors, such as computers and components (Table 6.4), the growth of employment in the industry as a whole has levelled off or slowed down in the 1970s — despite the continuing high rate of technical advance, the still rapid growth of sales and investment and the enormous future potential for applications of micro-electronics throughout the economy. Even in some sectors where demand is so strong that it has been little affected by medium and short-term business recessions in the 1970s, employment has tended to stagnate or even to fall, as in the case of the manufacture of telephone equipment. These tendencies have given rise to a widespread debate about the future implications for employment of micro-electronic technology. In the final section of this chapter we discuss these wider applications of the technology.

6.5 Micro-electronics

Almost every industrial and service activity is already affected to some extent by the microelectronic revolution, since all require information and in some of them, such as banking and publishing, information processing is the main function. The speed, scale and reliability of microelectronics mean that the automation and control of industrial processes in real time can be vastly extended. It is indeed difficult to think of an industry or occupation which will *not* be affected by microelectronics. The former Conservative Party spokesman on technology, Ian Lloyd, could think of only a few; among them were the makers of top hats, handloom weavers in the Outer Hebrides and psychoanalysts. He may well have been wrong about at least two of these.

Although 'automation' was by no means a new concept in the 1950s (since the thermostat is certainly a form of automation and was invented in 1625) it was the electronic computer which made it possible to realise in practice an enormous range of applications of the principle, which had been dreamt about before, but could not be successfully developed because of the limitations and costs of electro-mechanical technology. Thus, it is by no means unreasonable to link together the electronic computer, semiconductor technology, automation and information technology and to describe them collectively as a new industrial revolution or an 'information revolution'.

In fact, this group of inter-related technologies had already begun to revolutionize many industrial processes, office systems and weapon systems before the advent of the micro-processor. The significance of the micro-processor was that it drastically reduced the hardware cost of a technical revolution that was already well under way in the 1950s and 1960s. At the same time it both enhanced the performance of many existing electronic products and systems, and made it possible to introduce a great many new ones.

However, the fact that a new technology has many potential applications does not mean that all of these will occur simultaneously or even over a short time period. On the contrary we have argued that the assimilation of a new technology is a matter of decades rather than years, and in the case of the most important technologies may extend over a century, since a variety of enabling and facilitating social, managerial and educational changes may be necessary before the potential applications in many sectors can be realised in practice. This applies *a fortiori* to the international diffusion of new technological systems. We have already referred to the example of the railways and the automobile to make this point.

Table 6.4 Employment in the electronics industry (USA, UK, Germany (F.R.))*

		1958	1963	1967	1968
USA (SIC)					
3573	Electronic computing equipment	n.a.	n.a.	98.9	108.7
3651	Radio and TV receiving sets	66.5	81.3	116.7	112.5
3661	Telephone and telegraph appliances	80.1	89.5	115.4	117.0
3662	Radio and TV communication equipment	154.4	387.4	409.9	424.3
3671	Electron tubes, receiving type	37	25.9	21	18.7
3672	Cathode ray TV picture tubes	8.6	10.9	27.6	27.2
3673	Electron tubes, transmitting	20.1	18.7	18.2	18.7
3674	Semiconductors	23.4	56.3	85.4	87.4
3675-9	Electronic components, n.e.c.	108.9	176.7	251.3	235.8
367	Electronic components total	198.0	288.5	403.4	387.8
UK (MLH)					
366	Electronic computers	n.a.	10.4	n.a.	19.1
365	Broadcast receiving and sound reproducing equipment	n.a.	53.6	n.a.	41.6
363	Telegraph and telephone appliances	77.5	87.1	n.a.	87.9
364	Radio and electronic components	n.a.	84.7	n.a.	108.6
367	Radio, radar and electronic capital goods	n.a.	75.3	n.a.	78.9
Germany (F.R.) (IC)					
3651	Telephone and telegraph apparatus				
3653	and equipment				
3661	Radio and TV receiving apparatus and equipment				
3663	Sound reproducing apparatus				
3665	Electron tubes and semiconductors				
3667	Electronic components				
5050	Data processing apparatus and equipment				

*In thousands.

Sources: USA: *Annual Survey of Manufacturers*, various issues, and *US Industrial Outlook*, various issues; UK: *Census of Production*, various issues; Germany F. R.: *ZVEI, Statistische Mitteilugen Produktionsbericht*, various issues.

1970	1972	1974	1975	1976	1977	1978	1979	1980
145.7	144.6	177	163	166	193	231	267	306
90.0	86.5	88	69	72	75	91	87	84
155.6	134.4	145	119	105	124	130	135	141
389.7	319.2	318	316	316	335	372	406	415
14.3	11.4							
19.2	15.2							
16.8	20.5							
87.6	97.6							
217.0	190.9							
354.9	335.6	382	302	323	374	415	492	520
22.9	25.5	28.3	27.5	26.7	26.5	23.8	26.0	
52.0	48.5	54.1	50.5	45.1	43.3	38.9	36.7	
93.2	96.6	103.8	98.3	81.5	71.1	67.6	59.8	
126.2	125.4	135.0	116.8	109.7	111.7	109.9	107.4	
83.1	76.2	92.4	96.5	96.3	102.4	103.7	105.0	
	n.a.	98.2	92.0	84.0	84.1	88.1	96.5	
	n.a.	84.2	77.3	79.7	81.1	78.3	71.7	
	n.a.	33.4	31.1	28.8	27.0	26.95	25.6	
	n.a.	26.95	19.3	19.4	19.3	17.8	15.8	
	n.a.	53.9	52.2	50.0	46.9	51.8	49.4	
	n.a.	41.2	39.6	36.1	37.3	41.9	46.2	

In the case of electronics, visionary scientists and engineers, such as Wiener (1949) and Diebold (1952), were essentially right thirty years ago when they foresaw the many potential applications of electronic computers throughout the economic system in both factories and offices. Where they went badly wrong was in their estimation of the time scale. Wiener failed to take account of the long time lags in building up a capital goods supply industry and a component industry on a sufficient scale to provide all the computer power, peripherals and instrumentation for this vast transformation. Even more, perhaps, he underestimated the time scale needed to educate and train millions of people in the design, redesign, operation and maintenance of a huge variety of processes incorporating the new technology. Finally, he took insufficient account of the relative costs of the new technology which was still unattractive in purely economic terms for many potential applications.

As we have seen, during the 1950s, 1960s and 1970s there was indeed a massive expansion of the supply potential of the electronic capital goods and component industries and an enormous improvement in their relative costs and reliability compared with older electro-mechanical systems. But this still does not mean that there can be a sudden introduction of the new technologies throughout the economy, for several very important reasons. First of all, there is still a skill shortage especially in those sectors which have never had any wide experience of electronics technology. Secondly, the software costs in designing entirely new applications can be extremely high even though the hardware costs have been drastically reduced. Thirdly, full-scale automation may only be possible in association with other heavy re-equipment investment. Finally, in some important service sectors it may only be possible as a result of legislative, organisational, managerial and social changes which take a long time to bring about.

We have already suggested that the 'natural trajectory' of major all-pervasive new technologies, such as electronics, steam power or electricity, takes the form first of all of applications in the fastest growing newer industries where a lot of expansionary investment is in any case taking place, and where there is greater acceptance of innovations and greater availability of skills capable of using the new technologies. Thus the main beneficiary so far has been the electronics industry itself, which makes extensive use of computer aided design for both chips and end-products, of computer controlled production and stock control systems and of electronic office systems (Table 6.5).

Table 6.5 Diffusion of micro-electronic technology through the economy: An illustrative table. (This table is not intended to give precise data but to illustrate some of the main trends.)

Rate of Diffusion*	Rapid (from 1960)	Medium (from 1965)		Slow (from 1965)		
(Depth of impact)†	High	High	Medium	High	Medium	Low
Design and redesign of products to use micro-electronic technology	Electronic capital goods Military and space equipment Some electronic consumer goods	Machine tools Vehicles Electronic consumer goods Instruments Some toys	Other consumer durables Engines and motors Other machinery	Some biomedical products	Other toys	
Process automation using micro-electronic technology	Some electronic products	Machining (batch and mass) especially in vehicles, consumer durables and machinery Printing and publishing	Continuous flow processes already partly automated e.g.: • chemicals • metals • petroleum • gas • electricity	Clothing Textiles Food Assembly	Building materials Furniture Mining and quarries	Agriculture Hotels and restaurants Construction Personal services
Information systems and data processing	Specific government, business and professional systems involving heavy data storage and processing in large organisations	Financial services Communication systems Office systems and equipment without total electronic systems Design	Transport Wholesale distribution Public administration Large retailers	Retail distribution All-electronic office systems Electronic funds transfer	Domestic households Professional services	Agriculture Hotels and restaurants Construction Personal services

Source: Freeman (1982).
* Ranging from less than 10 years (rapid) to more than 30 years (slow) for the greater part of production to be affected.
† Proportion of total product or process equipment cost.

The second group of firms to make early and extensive use of the new technology were those where electronic subsystems represented a large part of total product cost and where the necessary skills were either already available (as sometimes in the case of scientific instruments and cash registers) or could be injected by aggressive strategies on the part of component suppliers (as in the case of calculators).

A third group of firms to use electronic technology rather quickly were those already operating large-scale flow production systems, especially those with innovative management facing expanding markets, as in the case of the chemical industry and the electric power industry in the 1950s and 1960s. Flow processes were already partly automated using older techniques. All of these three types of firm were able to achieve very big and sustained increases in labour productivity in the post-war period through the combined advantages of the new technology and the exploitation of scale economies (Chapter 7).

When, however, we come to consider the older and slower growing (or declining) sectors of the economy, especially those with a low endowment of qualified engineers, then the problems of diffusion are very different. Empirical diffusion studies suggest that thirty years is by no means uncommon for the time period over which revolutionary innovations are diffused through the majority of a potential adopter population and it can be much longer. These big variations in the responsiveness of different sectors of the economy to the diffusion of new technology are illustrated in Table 6.5. They account for much of the controversy about the impact of micro-electronics where those involved are often talking at cross purposes because they are referring to different sectors of the economy, where different time scales apply.

6.6 Summary

As in the case of plastics in Chapter 5, we shall now attempt to draw together this discussion of the introduction and widespread application of electronics technology and relate it to the more general issue of long cycles in the economy.

(i) Following the original growth of the radio industry in the early part of the century a series of inter-related revolutionary product and process innovations were introduced in the 1930s, 1940s, 1950s and 1960s. Clustering of innovations in the 1930s and 1940s was less clearly marked than in the case of plastics; innovations in solid state technology only started in the 1950s and continued for thirty years. Some important innovations (lasers, communication satellites, optical fibres, word processors, etc.) occurred only relatively recently. However, three of the most important of all innovations — television, radar and the computer — did occur just before, during and after World War II and were the main source of post-war growth of the electronics industry.

(ii) Again as in the case of plastics, all of these innovations and the revolutionary series of innovations in semiconductor technology owed a great deal to advances in fundamental science, in this case solid state physics. The development of both radar and computers depended on a very close and continuing interchange between the basic science community and industrial and military users. The wartime mobilization of physicists in the UK, USA and Germany greatly facilitated this interchange and the war generally acted as a forcing house for many new technical developments.

(iii) Although military demand was certainly important, normal commercial demand was much more sluggish and it was only after a fairly long period of science and technology 'push' that some of the most important commercial innovations got off the ground. A combination of government-sponsored, university and industrial R and D was important throughout, even after the swarming process was well under way in the 1940s and 1950s.

(iv) Once commercial applications did develop, the exceptionally large profits made by some of the innovating enterprises (especially IBM, RCA, Texas Instruments) did attract the Schumpeterian 'swarming' of imitators and an extraordinarily rapid growth process set in, associated with many further clusters of invention and innovation. While Schumpeter's mark II model represented much of this subsequent innovation fairly well, with established electronic firms playing an important role in using their in-house R and D facilities to keep pace with the technology and to advance it in some areas, there was also an important contribution from new innovative enterprises. This was especially important in the USA and in the components and instrument industries (see also Chapter 7). It was much less important in Europe and Japan.

(v) Even in areas where established oligopolies predominated, competition was sufficiently severe to erode profit margins fairly severely in the 1970s. An important feature was the severity of the international competition. At the same time economies of scale exerted an increasingly powerful influence and capital intensity increased rapidly. In the early phase of its growth in the 1940s and 1950s the industry had been extremely labour-intensive and took advantage of low labour costs in Asia and elsewhere, but this was changing by the 1970s. Consequently, although rapid growth of sales and investment continued throughout the 1970s the employment generating effects of the new technology diminished significantly.

(vi) There is still enormous potential for further growth and applications of micro-electronic technology throughout the economy but the time-scale for many of these applications will be a matter of decades and one of the main thrusts of new investment in the immediate future seems likely to be designed to exploit the labour-saving potential of the technology in other industries and services.

7 THE PATTERN OF POST-WAR STRUCTURAL CHANGE

The synthetic materials and electronics case studies of the two previous chapters illustrate well the big-boom phase of the post-war Kondratiev. The crucial new technology band wagons (drugs, plastics, television, transistors and computers, which were identified earlier on with the fourth Kondratiev upswing) all relate directly to both the chemicals and electronics industries. In Chapter 4, we argued that the impact of these new technology bandwagons was so important that they represented what we called 'new technology systems' rather than a set of haphazard bunches of discrete basic innovations and were, as such, directly responsible for elevating the whole economic system on to a higher level of economic growth.

In this chapter we will broaden the analysis to the more general issue of differences between the various sectors of the economy in their productivity growth, indicating the possible inter-sectoral differences in the rate of technical progress and differences between these various sectors in output growth, which reflect both technical change and changes in demand. In a first brief empirical section we discuss some of the most striking differences in economic growth between industrial sectors for a number of European countries over the post-war Kondratiev boom. In the second and third sections of this chapter we discuss, in more detail, the implications of the differential effect of technical progress on the various sectors of the economy: in other words the implications in terms of changes in industry structure, as well as changes in demand. In the final section we conclude our brief 'excursion' into dynamic demand issues, by relating it to both Schumpeter's mark I and mark II analysis.

7.1 A brief statistical exploration

From what has been said so far it is obvious that economic growth is far from a smooth, homogeneous process. Underlying overall macro-economic growth, one finds industrial sectors that are virtually disappearing and totally new sectors that are just emerging.

Following World War II many European countries achieved rapid economic growth in the first instance because of the initial massive capital destruction and the ensuing phenomenon of catching up. Yet, despite this generalized process of exceptionally rapid growth, which affected most sectors of the economy, growth rates differed substantially between the various industrial sectors. As illustrated in Tables 7.1 and 7.2 the fastest growth sectors (1960–73) were relatively 'technology-intensive' (e.g., chemicals, plastics, electrical and electronic equipment, instruments and communications) and industries meeting the changing pattern of consumer demand for goods with high income elasticities (e.g., cars, consumer durables, drugs, etc.). These sectors contributed significantly to overall economic growth in the various EEC countries, their output growth rate being nearly twice as high as the output growth rate for the whole economy. Interestingly, however, only in the case of the UK, the Netherlands and Belgium was the average growth rate of labour productivity in these sectors also nearly twice as high as the overall growth rate in labour productivity. As a consequence, the annual increase in job creation in these sectors was significantly higher than in the economy as a whole in most countries. Nevertheless, because of the greater gains in labour productivity and because of stronger pressures from international competition, the increase in prices in these sectors was markedly less than in the manufacturing sector generally or in the overall economy. The data shown in Table 7.2 are at a relatively high level of aggregation. More disaggregated statistics (e.g., for electronics within the electrical equipment industry and for plastics within the chemical industry) show even faster rates of productivity and output growth and still lower price increases for these subsectors.

By contrast and also illustrated in Tables 7.1 and 7.2 the declining sectors (agricultural products, textiles, clothing, shoes, solid fuels) had low output growth rates, well below the average for the economy as a whole. Yet while the growth in capital stock was also significantly below the overall growth rate, the growth in labour productivity was practically identical to the rate for the overall economy, resulting in considerable employment losses — excluding agriculture, the jobs lost in these sectors roughly equalled the jobs created in the growth sectors.

A similar picture to Table 7.1 and 7.2 could undoubtedly be drawn with respect to the USA, Japan and most other developed countries. Yet it must be stressed that the process of structural change can only be fully analysed at relatively high levels of disaggregation; this change can be at least as important within industrial sectors (e.g., from individual components to integrated circuits), as across the broad industrial sectors mentioned in Table 7.1. In

Table 7.1 Fast and slow growing sectors in the EEC, including UK: 1960–73

Germany	France	Italy	UK	Netherlands	Belgium
I. The fast growth sectors*					
Crude and refined oil, natural gas (22)	Chemical products (33)	Chemical products (33)	Minerals, building materials (32)	Chemical products (33)	Miscellaneous industry (64)
Chemical products (33)	Electrical equip. (44)	Motor vehicles and other means of transport (45, 46)	Chemical products (33)	Rubber, plastics (63)	Machinery, precision instruments and electrical equipment (42, 43, 44)
Precision instruments, data processing equipment (43)	Motor vehicles (45)	Ores, iron and steel (31)	Precision instruments, data processing equipment (43)	Crude and refined oil (22)	Chemical products (33)
Electrical equipment (44)	Agricultural and industrial machinery (42)	Miscellaneous manufacturing industry (64)	Electrical equipment (44)	Electricity, gas, water (23)	Rubber, plastics (63)
Rubber, plastics (63)	Communications (82)	Rubber, plastics (63)	Rubber, plastics (63)	Electrical equipment (44)	Crude and refined oil, natural gas (22)
Communications (82)	Crude and refined oil, natural gas (22)	Energy products (2)	Communications (82)	Ores, iron and steel (31)	Electricity, gas, water (23)
Banking, finance and insurance (84)	Electricity, gas, water (23)		Crude and refined oil, natural gas (22)	Motor vehicles (45)	Ores, iron and steel (31)
Electricity, gas, water (23)			Electricity, gas, water (23)		
Motor cars and other means of transport (45, 46)			Motor vehicles (45)		
II. The slow growth sectors*					
Agricultural products (1)	Agricultural products (1)	Agricultural products (1)	Ores, iron and steel (31)	Agricultural products (1)	Agricultural products (1)
Solid fuels (21)	Textile, leather clothing (61)	Building and engineering (7)	Metal products (42)	Textile, leather, clothing (61)	Textile, leather, clothing (61)
Metal products (41)	Solid fuels (21)		Textile, leather, clothing (61)	Solid fuels (21)	Solid fuels (21)
Agricultural and industrial machinery (42)			Solid fuels (21)		
Textile, leather, clothing (61)			Other means of transport (46)		
Hotels, catering (85)					

*Corresponding NACE-number between brackets.
Source: Adapted from EEC, *Maldague Report I* (1978).

Table 7.2 Characteristics of fast and slow growth industries (EEC: 1960–73)

Characteristic	Germany	France	Italy	UK	Netherlands	Belgium
Contribution of the fast growth sectors (in %) to total growth (in volume) of market economy	33.8	29.7	23.3	33.1	32.6	37.4
Average annual growth of value added (in %)						
Fast growth sector	7.2	9.0	7.6	5.3	11.9	9.3
Overall economy	4.2	6.0	5.3	2.7	5.2	4.8
Slow growth sector	1.9	2.3	1.8	−0.4	1.4	−0.3
Relative growth of capital stock (total economy = 100)						
Fast growth sector	130	125	n.a.	126	143	105
Total industry	123	102	n.a.	98	83	78
Slow growth sector	70	56	n.a.	64	n.a.	24
Average annual labour productivity growth (in %)						
Fast growth sector	5.1	6.5	5.9	4.7	8.6	7.6
Overall economy	4.6	5.4	6.1	2.7	4.3	4.2
Slow growth sector	4.7	6.0	6.1	2.3	5.2	3.9
Employment growth (in thousands)						
Fast growth sector	1087	648	322	278	96	114
Overall economy	−931	1267	−1620	199	368	123
Slow growth sector	−2015	−1877	−3152	−1088	−297	−260
Annual growth in prices (in %)						
Fast growth sector	2.9	2.5	3.6	2.6	1.3	1.3
Overall economy	4.3	4.4	5.4	4.5	5.0	3.8
Slow growth sector	3.4	4.6	6.5	4.3	3.5	4.5

Source: EEC, Maldague Report I (1978).

any case the process of differential *output* growth on the one hand and differential *productivity* growth on the other hand raises a number of important questions with respect to changes in the composition of demand, structural change and overall employment.

7.2 Structural change and technical progress

One of the main reasons why classical and neo-classical macro-economic growth theory (even when formally introducing technical progress) is of so little relevance to our long wave discussion, lies in the implicit assumption that technical progress (just like population growth) can be expressed in terms of an *overall* rate. This implies that not only is the rate of technical progress identical in all sectors of the economy, but also that demand growth is uniform across sectors. In such a world, 'full employment' is actually difficult to 'avoid' and even the attainment of a perfect 'natural' dynamic equilibrium growth path is pretty straightforward.* If this were true, economists would indeed be quite justified in spending most of their time on more important 'short-run' static problems.

As discussed earlier, in Chapter 4, technical change is not only unevenly spread over time, with swarms of innovations sweeping throughout the economic system in certain periods but not in others, it also has quite distinct influences on the various sectors of the economy — giving rise to fast productivity growth in some sectors and slow productivity growth in others, while at the same time generating whole 'new' sectors. This differential effect of technical progress in the various sectors of the economy is at the heart of economic change, to the extent that it leads to a continuous process of structural change within a wide range of fast and slow growing industries.

The crucial importance of technical progress in generating such a process of differential productivity growth was most clearly recognized by Salter (1960) who made it the centre piece of his analysis. While more attention is paid to the investment mechanism through which technical progress is transmitted to industrial practices in Chapter 8, here we will focus on Salter's description of the process of an initial uneven spread of technical progress as between industries leading directly to a process of differential productivity growth, which will be compensated in price adjustments to which demand in the form of output growth is assumed to respond directly. Salter's

---

*Each single sector and, by definition, the economic system as a whole expands at a rate equal to the rate of technical progress plus population growth.

view of the world was primarily supply oriented and stands in many ways in contrast to the more traditional demand determined models, such as that of Kaldor (1966), where productivity growth is not so much the result of the supply of technology, but rather is demand determined — with rapid output growth generating dynamic and static economies of scale and resulting directly in rapid productivity growth. While both theories each describe a different facet of the real world, for our present purposes we are more inclined to follow Salter's description to the extent that it puts the emphasis clearly on the rate of technical progress without ignoring economies of scale but rather assuming that they are secondary or in many ways dependent upon technical progress. Of course both views result in an identical 'hypothesis', namely that differential rates of productivity growth are closely and positively correlated to inter-industry output growth rates. The causality from one to the other is, of course, the opposite in the Salter case as compared to the Kaldor case.

Salter's empirical evidence for his 'model' was not only based on a close correlation between output and productivity growth, but also on the two crucial subresults implicit in his argument, namely that:

(a) inter-industry productivity growth rates were closely and negatively correlated to inter-industry price growth rates; and
(b) the latter were negatively correlated with output growth. In other words, in Salter's model productivity growth was associated with output growth *via* the influence of relative prices.

In updating Salter's analysis, which related to the period 1924 to 1950, Wragg and Robertson (1978) noticed that while the relationship between output and productivity growth was also clear for the post-war period, the relationship seemed to weaken for the period 1954-63, but strengthened somewhat for the more recent period 1963-73 — something for which we found confirmation in our own updating for the rather short period 1973-9* (see Table 7.3).

However, there remains for the post-war period a considerable

---

*An identical sample of eighty-two industries (as in the case of Wragg and Robertson) was used for the period 1973-9. Overall (i.e., in terms of the mean of all industries) there was a drop in average annual output growth over the period 1973-9 of 0.75 per cent, a fall in average annual employment growth of 2.12 per cent and an increase in average annual productivity growth of 1.38 per cent (all compound growth rates). This compares with average annual output growth rates of 2.5 per cent in 1954-63 and 3.2 per cent in 1963-73, average annual employment growth rates of −0.4 per cent in 1954-63 and of −1.2 per cent in 1963-73, and average annual productivity growth rates of 2.8 per cent in 1954-63 and of 4.2 per cent in 1963-73.

Table 7.3 Relationship between output per head (Y) and gross output (X)
(UK: 1924–79)

1924–50	$Y = 113.3 + 0.23X*$	$R^2 = 0.64$
1954–63	$Y = 91.0 + 0.31X*$	$R^2 = 0.39$
1963–73	$Y = 84.2 + 0.51X*$	$R^2 = 0.52$
1973–79	$Y = 11.5 + 0.64X*$	$R^2 = 0.50$

*Significant at the 1 per cent level.
Source: Wragg and Robertson (1978) for 1924–73; 1973–79 calculated from Census of Production.

amount of unexplained inter-industry variance in productivity and output growth. More specifically, more than a third of the industries analysed had either above-average output growth and below-average productivity growth or below-average output growth and above-average productivity growth. Wragg and Robertson them-selves advanced the Reddaway 'structural reorganization' hypothesis, which focuses specifically on the structural reorganization of de-clining industries (where the most inefficient plants or vintages are closed, leading directly to at least a measured increase in pro-ductivity of the remaining firms), as the most probable explanation. Indeed, a large number of industries with above-average, or rela-tively high, productivity growth but below-average, or sometimes negative, output growth were related to the textile and clothing industries.* An additional factor that might well explain the rapid productivity growth in these 'declining' industries is, of course, international competition, which (following Vernon's product-life cycle model) can be assumed to increase dramatically in the 'mature' phase of the industry.

As already illustrated in Section 7.1 the existence of declining industries, in which demand does not pick up whatever the in-fluence of productivity growth on relative prices,† is of course incompatible with the sort of simple supply-model described above. It is in relation to these industries, where the possibility of expand-ing demand is limited, that rapid technical progress will lead most directly to 'technological unemployment'. In many ways, one could

*With negative output growth rates but positive productivity growth rates: railway vehicles, coal-mining, hats, motor cycles, weaving, rope, spinning and doubling, jute, gloves, canvas, leather, woollen and worsted, textile furnishing, nuts and bolts (1954–73). It is interesting to note how in the more 'depressive' period 1973–9, this list nearly doubles to twenty-four industries (out of a total of eighty-two), including now (with above-average productivity growth, but below-average negative output growth), some of the major fourth Kondratiev growth industries such as office machinery, telegraph and telephone and man-made fibres as well.

†With price elasticities of demand well below 1.

identify 'depressions' as those periods in which the industrial structure is fundamentally unbalanced by an increasing number of industrial sectors entering their declining phase, and a decreasing number of expanding industries. The depression will last until sufficient capital is scrapped and redirected in the expanding sectors of the economy. It is best described as a prolonged period of creative capital destruction. Here, we are taking capital in its broadest sense; one of the most painstaking 'adjustments' probably relates to the 'scrapping of human capital' and the difficulties in generating quickly new skills. The fact that in some sectors in the economy (to some extent irrespective of the rate of technical progress itself, productivity growth or even pricing behaviour) output growth or demand flattens out after some time, points towards the crucial *demand* aspects of technical change and income growth.

7.3 Technical progress and changes in demand and industry structure

The overall employment effects of differential productivity growth depend in Salter's model on two crucial sets of elasticities — the price elasticity of output (or demand), and what could be called the 'productivity elasticity of prices' (i.e., the degree to which an increase in productivity growth leads to a proportional decline in prices). It should be stressed that productivity gains may not necessarily result in lower prices, but might be distributed in their totality to labour in the form of higher earnings, and/or to the capital owners in the form of higher profits. However, the most significant impact of productivity growth on *output* growth is through relative prices, where it is, in the first instance, the consumers who are the beneficiaries of the productivity gains. Therefore, it can be said that with sufficiently large price and productivity elasticities productivity growth should raise the growth of output enough to increase, or at least maintain, the level of employment.

As Wragg and Robertson noticed, while the evidence for the period analysed by Salter (1924-50) does indeed point towards such an effect, there is no longer any evidence for a close correlation between differential productivity growth and employment growth in the *manufacturing* sector for the post-war period. Confirmation of Wragg and Robertson's result is also found for the most recent period (1973-9) for which the sign of the relationship has actually turned negative (see Table 7.4).

Underlying this change, there is evidence of a further weakening in the relationship between output and employment growth, which

Table 7.4 Relationship between employment (Y) and output per head (X) (UK = 1924–79)

1924–50	$Y = 17.6 + 0.61X*$	$R^2 = 0.38$
1954–63	$Y = 71.7 + 0.19X$	$R^2 = 0.02$
1963–73	$Y = 75.9 + 0.08X$	$R^2 = 0.01$
1973–79	$Y = 10.1 - 0.07X$	$R^2 = 0.01$

*Significant at the 1 per cent level.
Source: as in Table 7.3.

had already been noticed by Cripps and Tarling and has been further confirmed by Wragg and Robertson's findings and our updating (see Table 7.5).

Table 7.5 Relationship between employment (Y) and output (X) (UK = 1924–79)

1924–50	$Y = 61.2 + 0.28X*$	$R^2 = 0.86$
1954–63	$Y = 33.8 + 0.50X*$	$R^2 = 0.73$
1963–73	$Y = 42.6 + 0.33X*$	$R^2 = 0.58$
1973–79	$Y = -10.0 + 0.42X*$	$R^2 = 0.39$

*Significant at the 1 per cent level.
Source: as in Table 7.3.

In other words, as compared to the inter-war period, employment creation has been less and less associated with the fast growing (or rapid productivity growth) sectors. This tendency was already noticed in both Chapters 5 and 6, where the employment generation of both the plastics industry and parts of the electronics industry fell quite substantially over the post-war period, particularly during the 1970s.

The change in the relationship between employment growth and both productivity and output growth over the post-war period seems to suggest that, as compared to the inter-war period, the price elasticity of demand and/or the productivity elasticity of prices have fallen. This is something that we would argue is directly related to the changing nature of industry structure, as well as to demand behaviour, over the various phases of the long wave. We will discuss changes in industry structure and relative prices first.

The process of rapid technical progress which accompanies the birth of new industries in the early phases of a long wave, gives rise both to the emergence of new firms (growing at phenomenally high rates and leading after a number of years to rather important changes in the overall industrial structure and the concentration of the economy as in Schumpeter mark I), and to the rapid growth of those branches of large firms which promote the new technology (as in Schumpeter mark II).

With respect to post-war economic growth one has only to refer to the phenomenal growth rates of such firms as IBM (mark II), Xerox and Polaroid (mark I) in the 1950s and 1960s. This has not only lead to crucial changes in the overall structure of the economy of the USA, as a glance at the 1950 and 1970 Fortune 500 list will reveal, but it has also, and perhaps paradoxically, been accompanied by a significant increase in overall industrial concentration. Table 7.6 contains some information on overall concentration for the USA and the UK based on census data. While the increase in overall concentration is particularly striking in the case of the UK, the 14 per cent increase in the share of total output accounted for by the 200 largest US companies over the period 1947–76 is also impressive.

The tendency towards increased industrial concentration is in our view a typical characteristic of the 'prosperity–maturity' phase of the long wave. In the early phases, or in the final phases of the previous Kondratiev, the structure of the new industries (which will eventually become the major 'Kondratiev-carriers'), is highly unstable; the innovating firm(s) may be forced to take substantial risks, with very little scope* for preventing imitators from entering and taking over, thereby causing the innovator to fail.

Over time, however, with the maturing of the industry, entry is not only becoming more and more difficult, it is also becoming less and less attractive for potential entrants as the process of the 'competing away' of profits gains momentum. With the need for further cost reductions, the pressure for standardization (needed to realise the crucial economies of scale) increases, and the industry becomes gradually more concentrated with entry becoming effectively deterred. In addition, the devotion of large resources to R and D should at least allow the firm to play the role of the fast imitator without having to incur most of the risks, whenever a radically new innovation is brought on the market by a new firm. When the latter becomes successful against all the odds there remains even the relatively costly option for the established firm to acquire the new firm.

In the course of the gradual change in the structure of an industry

---

*Legal protection under the form of patents will always be sought, yet it will not always be a sufficient guarantee for preventing imitation. This could well explain the difference in the propensity to patent between large and small firms (Soete 1979b).

Table 7.6 Industrial concentration in the USA and the UK (in percentages)

USA						
Share of total value added by	1947	1954	1963	1967	1972	1976
The largest 50 companies	17	23	25	25	25	24
The largest 100 companies	23	30	33	33	33	34
The largest 150 companies	27	34	37	38	39	40
The largest 200 companies	30	37	41	42	43	44
UK						
Share of total net output by	1949	1953	1963	1968	1972	1978
The largest 100 companies	22	27	37	41	39	39

Source: USA: *US Department of Commerce* (1980); UK: Prais (1976) and *Census of Production* (1972, 1978).

from its birth to its decline, it seems possible to identify a shift in the industry's pricing behaviour from a high unstable initial mono- poly one — in the very early phases, where one might well assume that most of the productivity gains will accrue to the *capital owners*, under the form of 'pure' monopoly profits — to a more competitive one in the still early phases of the industry's life cycle, with some- times a dramatic fall in prices. Later a more 'oligopolistic' pricing behaviour develops in the more mature phases of the industry's life cycle when there is either price-setting by the leading firm or a more elaborate price cartel agreement.

The gradual shift in the early phase from the innovator's initial monopoly to a more (and sometimes extremely) competitive situa- tion depends, of course, to a crucial extent on whether the innova- ting firm fails in solely 'appropriating' its own innovation. Established innovators (mark II) might often be more successful in preventing imitation than the new innovating firms (mark I). In addition, in some industries (e.g., the drug industry with librium and valium), innovators might be able to use patent protection as a sufficient technological entry barrier and continue to enjoy 'pure' monopoly profits over a relatively long period.

In most other industries, however, it is the entry of imitators over the fast-growing phase of the industry or product life cycle that leads to the 'competing away' of these profits, a fall in prices, and a shift of the gains from productivity away from the capital owners to the consumers in the form of lower prices. This shift, while reducing the attraction of entry to other firms, is important for the future growth of the industry, to the extent that it provides a direct and essential 'impulse' to the industry's output growth. Once that growth is clearly identified, firms will try to use static and dynamic economies of scale to the best of their advantage. There will be a gradual shift from a relatively 'open' price competition structure to a more oligopolistic 'closed' pricing behaviour. To the extent that oligopolistic pricing deters potential new entrants sufficiently, firms might become more ready to distribute the productivity gains to their *employees*, having accounted for a sufficient profit-margin. From a single firm's point of view the distribution of the productivity gains amongst its own workers is actually more advantageous than the 'impersonal' spreading of these gains over all consumers. Not only does it allow firms to 'buy' social peace from its workers and trade unions, it also, as Pasinetti (1981) observes, creates a 'labour aristocracy', with relatively weak loyalties to the trade unions. It also establishes stronger links between workers and 'management', where the workers' interests can be directly identified with the firm's interests.

What might upset this relatively stable 'oligopolistic' industry most directly is *foreign* competition, based on some absolute cost advantage. Again the productivity gains, here primarily trade gains, will accrue primarily to consumers through lower prices. Downward wage inflexibility will normally lead to exchange rate adjustments or increased pressure for protectionism, leading to possible further reprisals, etc. In the long run the most probable outcome is, however, some 'new' form of international oligopoly structure, with agreed international price setting, 'voluntary' export restraint, etc.

In relation to the specific question of why the 'productivity elasticities of prices' might have fallen in the manufacturing sector over the post-war period, the previous discussion suggests that over the fourth Kondratiev long wave, and with the flattening out of growth — at least with the 'maturing' of some of the crucial main Kondratiev carrier industries — the structure of these industries (as well as overall industrial structure) has become more concentrated, with firms having become gradually more ready to distribute the productivity gains to their employees rather than to the consumers in general under the form of lower prices. It is also worthwhile noting that to the extent that such a policy prevailed throughout most sectors in most Western economies in the 1970s, it might well explain the phenomenon of widespread creeping inflation in terms of an upward adjustment of prices in those sectors with below average productivity growth.

As Pasinetti observes, if the sectors with rapid productivity growth do fail to cut prices in any significant way, the only way for the price system to restore or maintain its allocative efficiency is through a corresponding rise in the general level of prices. If one fails to observe this price adjustment mechanism, any attempt to limit the overall price level from increasing based on purely monetary considerations, will in the long run lead to further distortions in the price system. While these attempts might prove relative successful in the short run:

they can only succeed to the extent that they make a price distortion permanent, thereby generating inefficiencies, or, even worse, to the extent that, by discouraging investments, they slow down the whole process of increase in productivity and therefore of economic growth. (Pasinetti 1981, pp. 221–222).*

*This is actually directly admitted within present British monetary policy: 'A constraint on the growth of money expenditure implies that an increase in investment, whether public or private, must be accompanied by some reduction in other expenditure'. In other words any upswing of 'autonomous' investment, resulting, for example, from a major product innovation, will result in government steps to reduce expenditure elsewhere in the economy by the same amount. How an economy might 'recover' from a severe recession/

The second factor which might explain the further weakening of the relationship between employment, output and productivity growth over the post-war period, relates directly to a possible fall in the price elasticity of demand. Indeed, the possibility of demand becoming more influenced by income elasticities and less by price elasticities, as mentioned by Wragg and Robertson (1978) amongst others, provides an additional incentive for firms to distribute their productivity gains under the form of higher earnings rather than lower prices.

The whole issue of the evolution of demand over time (in particular its composition) has received little attention in economic theory. The only dynamic demand theory that has been subject to empirical testing and has received 'law status', is the rather old 'Engel Law', which states that the proportion of income spent on food declines as income increases. Crude generalizations of this law have undoubtedly been responsible for a number of general consumer demand saturation arguments, which though quite valid for a number of individual durable consumer goods (in particular the so-called white goods such as refrigerators, cookers and washing machines) are not directly supported by the overall evidence on the 'marginal propensities to consume' of most Western economies, as these do not seem to have fallen (or increased) in any significant way during the 1960s or 1970s.

A more correct general formulation suggested by Pasinetti (1981) of Engel's Law points, however, to the 'non-uniform' and 'non-proportional expansion' of demand with increase in income; more specifically he states that 'the proportion of income spent *on any type of good* changes as per capita income increases' (p. 70). To a large extent these changes will occur *independently* of the price elasticity of demand for these goods, but according to some sort of *order of priority* of consumer needs. Prices, while of crucial importance at any given level of income (i.e., within primarily

depression under such a policy where increases in *investments* and other components of demand will be matched by comparable reductions elsewhere in the economy, is of course the crucial question. According to H.M. Treasury:

The answer is that a reduction in the rate of inflation is a pre-condition to the resumption of economic growth, because as inflation falls the growth of nominal expenditure permitted by our existing monetary and fiscal policies will increasingly consist of a rise in real output, and that is how we will grow (H. C. 348 HMSO 1981, as quoted in Kaletsky 1981)

If, however, as Pasinetti observes, inflation is primarily the response of the economic system to restore price efficiency (compare, e.g., the price rises in some of the nationalized 'declining industries', with overall price rises), the only way the economy can 'recover' or 'resume' economic growth is through a different economic strategy.

a static framework), 'can only postpone or anticipate a time path which, if real income increases is going to take place anyhow'.

In a situation of rapid income per capita growth such as occurred in the 1960s and early 1970s, there is little doubt — from a purely theoretical perspective — that price elasticities might have fallen, and that demand itself would become more and more dependent on income elasticities. With rapid income per capita growth it is the development of new products corresponding to new consumer wants that is the crucial factor in maintaining the balance between the rate of growth of productivity and the rate of growth of output, and by implication full employment.

7.4 Technical change, recovery and the identification of new demand

The emphasis put on the changing nature of 'demand' and its relationship with technical progress in the previous section, leads to one of the major demand explanations for recessions or slumps, directly related to the differential 'structural' effects of technical change on demand. It can best be described in terms of more or less typical 'Keynesian' demand deficiency periods — more precisely, periods in which the growth rate of effective demand falls short of the rate of growth of the production potential because of increasing effects of saturation in a large number of existing mature goods and because of a failure to identify new consumer wants.

Again, Pasinetti (1981) has provided an interesting elaboration of this Keynesian interpretation of recessions or slumps. In his view, Keynesian demand-stimulating policies, while useful in terms of avoiding some of the *negative* aspects of recessions, will not alter the 'structural demand' problems, which are related to the fundamental difficulty of the economic system in identifying new consumer wants, and they are therefore unable to cause a resumption of economic growth.*

---

*According to Pasinetti:

'When per capita incomes are increasing, consumers are pushed into new and previously unexperienced fields of consumption, in which they are compelled to make their choices; and it may from time to time become very difficult for the investors to detect clearly which are the directions in which consumers' choices are going to expand (or may be induced to expand). When this happens, and some uncertainty falls over the future direction of expansion, *the simple attitude of waiting* or of postponing the actual undertaking of investment, will cause the weighted sum of the investment coefficients to fall . . . . Unemployment will inevitably appear, and it will be unemployment *both* of productive capacity *and* of labour force.' (Pasinetti 1981, pp. 233-4).

'Had this been correctly understood many recent disillusionments with Keynesian theories

According to Pasinetti, the latter can only be overcome by *speeding up the rate of learning*,* something which the multi-product large firms might be particularly good at, 'by keeping backlogs of ideas to be used when needed' (p. 235). Pasinetti's definiton of the 'rate of learning' is unfortunately neither clear nor useful when describing how the system eventually moves out of its self-imposed 'pause'.

The most crucial factor affecting this rate is, of course, technical progress itself, which through the development of primarily new products and through process innovations will enable effective demand to 'recover' — either by allowing new consumer wants to be satisfied through new or improved products, or by broadening the scope of consumer satisfaction of existing goods to lower income classes.

In Pasinetti's system, technical progress, while fully differentiated between sectors, is still chiefly considered as taking place in a more or less continuous way,† whereas in our system technical change is fundamentally discontinuous.

In addition, the question of whether the multi-product large firm will necessarily be the most 'helpful' or efficient in identifying

---

and policies might have been avoided. We cannot expect from the Keynesian theories and policies what they cannot give. We have gained from them the avoidance of large scale unemployment and this has been a notable achievement. But the resumption of growth is another matter. The economic system still has to solve the much deeper problems . . . the structural problems of learning the appropriate ways to expand.' (Pasinetti 1981, p. 238).

---

*'The very nature of the process of structural growth — i.e., of consumers' preferences and of producers' organizational requirements — *imposes periodically* on the economic system, both on the part of consumers and on the part of producers, the necessity of *speeding up the rate of learning*, if unemployment is to be avoided. Unfortunately the rate of learning is not something which can be manipulated at will. Here again multi-product large firms may be of help in increasing the stability of the economic system, by keeping backlogs of ideas to be used when needed. However, the natural consequence will clearly be that, periodically, the direction of expansion will indeed become rather uncertain, that new outlets will not immediately be at hand and that the rate of investment will inevitably flatten down . . . . Some unemployment will begin to appear and the economic system will at least enter a period so to speak, 'of pause', until new types of products for expansion or new interests in old products are found (which is a learning process that may take quite some time). Only then will expansion be started all over again.' (Pasinetti 1981, p. 235).

---

†'Technical progress is a very complex phenomenon . . . it includes all the innumerable series of expedients and devices, small if considered individually, but of great relevance if jointly gauged, which are the daily upshot of experience, experiment, research and rethinking of the organisation of production. This is indeed a very complicated process emerging from the learning activity of human beings and the application of this learning activity to production. By its nature this process is therefore a slow, but persistent, one.' (Pasinetti 1981, pp. 66–7).

'new consumer wants', thus enabling the system to speed up its 'state of learning', depends in the first instance on its technological commitment, which might well be in a large number of relatively mature consumer wants.

Pasinetti's concept of the multi-product large firm is of course most closely comparable to what we typified as the Schumpeter mark II model, with the incorporation in large firms of most scientific and technological activities. While we have argued earlier on (Chapter 2), that the mark II model described the inter-war period and most recent period (the 1970s) well, we also stressed the fact that long-term cyclical upswings might be more closely associated with a resurgence of mark I small firm innovators.

The important question is, indeed, whether it will be the small highly innovative firms, rather than the multi-product large firms, which will be ready to take the risk and enter the new, uncertain markets. While the large multi-product firm will no doubt diversify into new areas, most of its output will be into relatively mature industries or products. Its commitment to the new areas might, therefore, not be total and parts of the profits earned within the new fast-growing sectors might well be used to cover possible losses in the established maturing or declining sectors, rather than be reinvested in the new sectors. A clear picture emerges out of the growth of firms in the USA in the most recent relatively 'recessive' period of 1975–80, which is most clearly comparable to Pasinetti's 'pause' period. In Table 7.7, we have listed *all* US firms which had average *annual* growth rates in sales of more than 40 per cent over the period 1976–80 from a total sample of 750 R and D performing firms, reported in *Business Week* (June 6th 1981).

Apart from the fact that, as expected, most of the firms listed in Table 7.7 are relatively small, the most striking feature of Table 7.7 is that these are all firms within the broad technological area of information processing and electronics, which has often been heralded as the new technological cluster around which a major new upswing might be set in motion. Out of the thirty-six firms, twenty-eight fall either within information processing (computers six, office equipment three, and services eleven), electronics (five), semiconductors (two) or telecommunications (one), with the rest of the firms in instruments (three), miscellaneous (three) and machinery (one). The list includes some of the most well-known new names in 'home computers' such as Tandem and Apple Computers, in CAD, such as Computervision and in 'office equipment', such as Wang Laboratories. Overall, their average annual growth in profits has been as impressive as their output growth and even

Table 7.7 Firms with annual average growth rates in the USA (sales 1976–80) of more than 40 per cent

Firm	Sector	1976–80 Average annual growth			1980		
		Sales (in %)	Profits (in %)	Empl. (in %)	Sales (million $)	R and D/ Sales ratio (in %)	Profits/sales ratio (in %)
Tandem Computers	Information processing: computers	247.4	284.5	n.a.	109	8.1	10.1
Cray Research	Information processing: computers	197.0	126.8	71.8	61	14.5	18.0
Apple Computer*	Information processing: computers	144.7*	n.a.	11.2	117	6.2	10.3
Floating Point Systems	Information processing: computers	120.5	67.0	n.a.	42	10.9	9.5
Intermedics	Electronics	111.5	113.4	53.3	105	4.6	10.5
Triad Systems	Information processing: peripherals services	89.5	110.3	n.a.	57	6.6	8.8
Prime Computers	Information processing: computers	88.1	132.0	73.8	268	7.6	11.6
Rolm	Telecommunications	79.0	101.4	69.7	201	6.7	8.5
Lamson & Sessions	Miscellaneous manufacturing	65.3	n.a.	26.8	235	0.6	–1.3
Auto-Trol Technology	Information processing: peripherals services	64.1	96.9	n.a.	51	12.1	7.8
Data Terminal Systems	Information processing: peripheral services	64.0	n.a.	49.5	118	5.1	–2.5
Computervision	Information processing: peripherals services	59.0	116.3	44.7	224	9.9	10.3
Paradyne	Electronics	57.9	132.8	51.8	76	8.4	10.5
Siltec	Electronics	56.5	64.0	n.a.	57	3.6	5.3
Advanced Micro Devices	Semiconductors	56.2	91.0	48.6	226	12.5	10.2

Savin	Information processing; office equipment	53.7	90.2	33.5	357	2.3	7.8
American Management Syst.	Information processing; peripherals services	53.3	57.2	n.a.	59	7.1	3.4
CPT	Information processing; office equipment	49.4	55.5	37.9	59	3.4	10.2
Wang Laboratories	Information processing; office equipment	48.5	72.1	n.a.	543	6.7	9.6
Storage Technology	Information processing; peripherals, services	48.2	58.0	n.a.	604	6.5	7.5
Comshare	Information processing; peripherals, services	47.6	51.4	38.9	78	6.3	5.1
Datapoint	Information processing; peripherals, services	47.3	61.8	36.6	50	8.7	10.3
Verbatim	Information processing; peripherals, services	47.3	28.5	45.6	319	5.8	2.0
US Surgical Instruments	Instruments	46.9	52.8	34.1	86	3.5	9.3
Sega Enterprises	Miscellaneous manufacturing	46.8	55.5	9.3	140	1.2	8.6
Parker Pen	Miscellaneous manufacturing	45.2	34.8	11.2	664	0.4	6.0
TIE/Communications	Electronics	44.9	94.3	n.a.	60	2.5	5.0
Kratos	Instruments	44.0	44.5	n.a.	56	7.5	5.4
Intel	Semiconductors	43.9	43.4	27.5	855	11.3	11.3
Datacard	Information processing; peripherals, services	42.6	55.5	n.a.	66	2.6	10.6
Gerber Scientific	Instruments	41.6	83.9	32.1	74	7.1	8.1
Data General	Information processing; computers	41.0	33.0	31.0	654	10.0	8.4
Computer Consoles	Information processing; peripherals, services	40.7	55.8	16.1	44	10.5	11.4
Analogic	Electronics	40.3	67.7	n.a.	67	8.9	9.0
Miller (Herman)	Miscellaneous manufacturing	40.3	40.0	35.1	230	2.5	5.2
Pengo Industries	Machinery	40.1	22.8	n.a.	78	1.9	3.8

*For Apple Computer Inc. which only became publicly held in 1980, the growth in sales relates only to the period 1979–80.
Source: Calculated from Business Week, June 6th 1981.

their employment creation appears, from the information available, to have been rather impressive.

All these firms, with the exception of the machinery firm and the few miscellaneous firms, are what could be called small highly innovative firms, which are ready to take substantial risks to enter new markets. It is interesting to compare the performance of these firms with the performance of the major established firms in these same sectors (see Table 7.8).

Of the firms listed in Table 7.8 only five (Control Data, Hewlett-Packard, Honeywell, Motorola and Texas Instruments), had annual growth rates in sales above the average for all the firms in the sample (i.e., 16.6 per cent). As indicated by the R and D/sales ratio, most of these firms devoted large parts of their sales to R and D; on average only slightly less (5.5 per cent) than the fast-growing firms listed in the previous Table (6.5 per cent), yet their profit/sales ratio (5.7 per cent) was well below the average profit/sales ratio (8.3 per cent) of the firms listed in Table 7.7. While there is little doubt that these firms will also move quickly into uncertain new markets, their commitment to the production/servicing of more mature products (large computers, large office equipment, electronics calculators and watches etc.) might not allow them to expand as rapidly into the fast growing new demand areas as the newcomers.

It is, paradoxically, the dual relationship between innovation and market structure that is at the core of the slowing down and speeding up of the identification of 'new consumer wants'; this is Pasinetti's so-called 'learning'. As described earlier, throughout the growth of an industry, when effective demand is high and 'well-defined', technical progress itself will slowly 'mould' the oligopolistic structure of the industry; however, this structure, in turn, influences the industry's innovation strategy, and might even in some cases prevent the established firms from identifying new consumer wants. As Schumpeter says:

However high barriers to entry may appear to be, they offer the firms who have built them *no* definite guarantee gainst the type of competition based on innovation which strikes at the foundations of existing industries and of the firms engaged (Schumpeter 1939, see Fels (1964), p. 84).

Table 7.8 US established firms in the electronics/information industries

Firm	Sector	1976–80 Average annual growth			1980		
		Sales (in %)	Profits (in %)	Empl. (in %)	Sales (million $)	R and D/Sales ratio (in %)	Profits/Sales ratio (in %)
North American Philips	Electronics	12.8	16.6	6.0	2658	1.4	2.8
RCA	Electronics	11.1	21.4	3.3	8011	2.5	3.9
Raytheon	Electronics	14.9	27.7	6.1	5002	2.6	5.6
Burroughs	Information processing: computers	12.1	–5.1	2.8	2857	6.8	2.9
Control Data	Computers	18.3	34.1	10.1	2766	6.6	5.4
Hewlett-Packard	Computers	26.4	27.3	14.9	3099	8.8	8.7
Honeywell	Computers	17.3	30.3	8.3	4925	6.0	5.7
IBM	Computers	12.6	11.2	n.a.	26213	5.8	5.8
NCR	Computers	10.9	41.3	1.7	3322	6.1	7.7
Sperry	Computers	12.7	20.5	n.a.	5427	6.2	5.8
Rank Xerox	Information processing: office equipment	15.0	13.2	4.5	8197	5.3	7.6
Motorola	Semiconductors	19.0	28.5	9.3	3099	6.5	6.0
Texas Instruments	Semiconductors	24.5	25.8	9.6	4075	4.6	5.2

Source: as in Table 7.7.

8 SOME STATISTICAL CHARACTERISTICS OF THE 'FOURTH LONG WAVE'

8.1 Introduction

In this chapter some macro-level data on post-war economic growth are presented. The purpose is to discuss the extent to which they provide support for the characteristics of different phases of the 'long wave' suggested in earlier chapters.

The pattern of economic development since World War II has in many respects been historically unique. Rates of GDP growth during the prosperity phase of the fourth Kondratiev (Table 8.1)

Table 8.1 Average annual growth rates of gross domestic product

	1870–1913	1913–50	1950–60	1960–70	1970–80	1973–80
France	1.7	1.0	4.7	5.6	3.5	2.8
Germany*	2.8	1.3	8.1	4.8	2.8	2.4
Italy	1.5	1.4	5.1	5.3	3.1	2.8
Japan	2.5	1.8	8.6	10.3	4.7	3.2
UK	1.9	1.3	2.7	2.7	1.8	1.0
USA	4.1	2.8	3.2	4.2	2.9	2.1

Sources: Maddison (1979); Brown and Sheriff (1979); OECD (1981).

were high, even in comparison with earlier prosperity phases, as was the growth rate of labour productivity (GDP per manhour, see Table 8.2). As these tables indicate, there were considerable

Table 8.2 Average annual growth rates of labour productivity (GDP per man-hour)

	1870–1913	1913–50	1950–60	1960–70	1970–80	1973–80
France	1.8	1.7	4.3	5.1	3.8	3.7
Germany*	1.9	1.2	6.6	5.2	3.6	3.2
Italy	1.2	1.8	4.3	6.3	2.5	1.7
Japan	1.8	1.4	5.7	9.6	4.3	2.6
UK	1.1	1.5	2.3	3.2	2.4	1.6
USA	2.1	2.5	2.4	2.4	1.5	0.8

*Federal Republic 1950–80
Sources: 1870–1950; Maddison (1979, 1980); 1950–80: OECD for GDP and Employment, ILO (1980) for working hours.

discrepancies between countries regarding economic performance, both during the post-war 'prosperity' phase and during subsequent recession (and depression) phases. The stronger economies, notably Japan and West Germany, have (at least to date) suffered relatively less deterioration in economic performance, growth rates in these countries having remained high by historical standards. As noted in Chapter 1, however, these countries have not been entirely free from pressures tending to increase unemployment.

Tables 8.3 and 8.4 (from Maddison 1979) show long-term trends in the growth rates of capital stock and the capital–labour ratio, respectively. While the problems involved in compiling statistics

Table 8.3 Rate of growth of total non-residential fixed capital stock (Annual average compound growth rate)

	1870–1913	1913–50	1950–70	1970–7
France‡	n.a.	(1.1)	5.4	6.3
Germany	(3.1)	(1.0)	6.2	4.8
Italy	[2.5]*	[2.2]	[5.1]	[5.0]
Japan	2.7†	[3.3]	8.8¶	7.9¶**
UK	1.4	0.7	3.9	3.7
USA	4.7	2.0	3.8	3.0
Arithmetic average	2.9	1.8	5.5	5.1

Notes: see footnote to Table 8.4.
Source: Maddison (1979).

Table 8.4 Rate of growth of non-residential fixed capital stock per man hour (Annual average compound growth rate)

	1870–1913	1913–50	1950–70	1970–7
France‡	n.a.	(1.8)	5.2	8.0
Germany	(2.1)	(0.9)	5.9	7.1
Italy	[2.3]*	[2.6]	[4.9]	[7.3]
Japan	2.0†	[2.9]	6.8¶	8.4¶**
UK	0.6	0.8	4.0	4.4
USA	2.6	1.8	2.7	1.8
Arithmetic average	1.9	1.8	4.7	5.7

Notes: The figures in Tables 8.3 and 8.4 are adjusted for geographic change. Except for the bracketed figures, these data represent an average of gross and net stocks; gross stocks are calculated on the basis that an asset retains its full value during its working lifetime, while in calculating net stocks allowance is made for depreciation. Figures in round brackets refer to net stock only, figures in square brackets to gross stock only.
*1882–1913; †1880–1913; ‡ refers to private stock; ¶ net stock refers only to the private sector; **1970–6.
Source: Maddison (1979).

such as these are particularly severe, the impression here also is of unprecedently high growth rates since 1950, with the exception of the USA. Overall, there is no evidence of a significant decline in the growth of fixed capital in the 1970-7 period compared with the previous two decades, while growth in capital-intensity appears to have accelerated. Maddison (1979) concludes that 'the fundamental instrument (on the supply side) for faster post-war productivity growth has been the acceleration in growth of the capital stock per hour worked'; and a favoured explanation for the post-1973 productivity slowdown, particularly in the USA, is in terms of a reduction in the rate of capital accumulation — although no wholly satisfactory explanation for this slowdown has been provided (see e.g., OECD 1980). Trends in investment in fixed capital, and its relationship to technical change and employment, are discussed in later sections of this chapter.

In the fourth long wave the recession period has been characterized by high and growing rates of inflation (Table 8.5). This feature, as indicated in Chapter 1, is central to some recent interpretations on unemployment. While our primary interest here is on variations in 'real' indicators, these are not, in any view, unrelated to inflation. For example, there seems to be little doubt that uncertainties about inflation can adversely affect general expectations about the future, and that attempts to overcome inflation by government policy are counter-expansionary.

As mentioned in Chapter 7, it is noteworthy that rising prices may be expected in an environment in which the Schumpeter mark II model, described in Chapter 2, is applicable. A tendency towards increasing industrial concentration and oligopoly may be expected to be accompanied by a reduction in the downward flexibility of prices. When combined with a readiness of firms to offer high rewards for scarce skills, and an ability of strong unions to maintain differentials, forces leading to a wage–price spiral are present. Increases in raw material (particularly oil) prices have certainly exacerbated this process, although it is noteworthy that inflation was becoming significant in most countries well before the OPEC price increases of 1973.

Two quantities often assumed to remain constant in formal growth-theory models — the productivity of capital and the shares of total output accruing to labour and to profits — have undergone apparently significant shifts over the last three decades. Figure 8.1 shows the trend in the productivity of capital (i.e., the output–capital ratio) for four countries, while Figure 8.2 shows the (gross) rate of profit, equal by definition to the profit share

Table 8.5 Consumer prices: Annual percentage increases, in national currencies

	1960–73 (average)	1974	1976	1978	1980
West Germany	3.7	7.1	4.5	2.5	5.5
France	4.7	13.2	9.8	8.5	13.6
Italy	4.8	20.9	18.1	12.8	21.2
UK	4.9	17.3	15.4	8.6	18.4
(Total EEC)	4.5	12.7	10.3	7.3	12.0
USA	3.2	11.0	5.8	7.7	13.5

Sources: *European Economic Community, Annual Economic Review 1980–1* (November 1980) No. 7, Table 2.1; US Department of Commerce (1980); OECD Main Economic Indicators, OECD, Paris, June 1981.

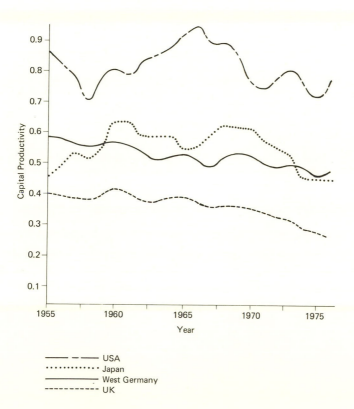

Figure 8.1 Capital productivity (output–capital ratios), 1955–75. *Source*: Drawn from the data estimated by Hill (1979).

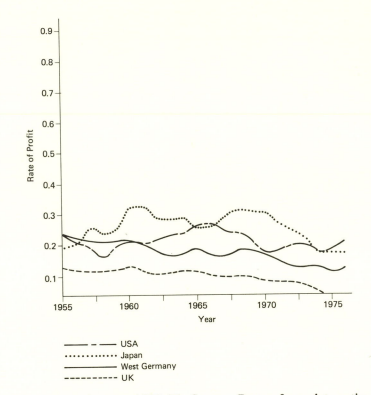

Figure 8.2 Rates of return 1955–75. *Source*: Drawn from data estimated by Hill (1979).

multiplied by the capital productivity. There seems little doubt that the rate of profit has declined in most countries since the late 1960s (the USA appears to be an exception); and further, that this decline derives partly from a reduction in the share of profits and partly from a decline in capital productivity. These trends are by no means universal, and measurement problems are severe — see Hill (1979), from which these graphs are derived, for a discussion of the methodological difficulties. The overall picture, is, however, highly suggestive of a general decline in rates of return in Western European economies.

8.2 Employment and output

Some evidence for the implications for employment of post-war trends in output and productivity, as conditioned by the developments referred to above, is provided by Figures 8.3, 8.4 and 8.5. Attention is drawn at the outset to the different scales used in the axes of these figures — for present purposes it is the overall trends which are of interest, rather than absolute values. Figure 8.3 shows industrial employment plotted against industrial output for the nine EEC countries and Figures 8.4 and 8.5 show manufacturing output and employment for the USA and Japan respectively, using index numbers and taking 1962 as equal to 100 in each case. While output growth since 1962 has been broadly similar in the USA and EEC, having roughly doubled in each case, employment trends show markedly different behaviour in these regions. For the combined EEC countries, Figure 8.3 — an updated version of that discussed in Soete (1978) — suggests that the period can be separated into three rather clear phases: one in which both industrial output and industrial employment were expanding fairly steadily (1950–66 approximately); a second (1961–74) in which output continues to grow rapidly but employment is static; and a third, in which output growth is very slow and in which employment declines. Of particular interest is the fact that the growth of labour productivity (output per man) 'caught up' with the growth of output during the 1960s, giving a period of 'jobless growth' despite the continuation of (and even slight increase in) the high growth rate of industrial output. This suggests a shift in the relationship between output and employment; in particular, that there has been an increase in the rate of growth of output needed to sustain a given level of employment (or, equivalently, an increase in the 'underlying' growth of productivity associated with a given output growth). Direct attempts to establish the evidence for such a shift from estimation of employment–output and productivity–output relationships (e.g., Clark, Freeman and Soete 1981a) suggest that, in some economies at least, significant shifts have occurred — although the picture is by no means totally clear cut and problems of interpretation are considerable. What can certainly be claimed is that, where a significant shift has occurred it is in the direction of 'rationalization', whereby a given level of output growth is increasingly achieved by productivity growth rather than by employment growth. This trend is consistent with standardization and the increasing availability and exploitation of scale economies. In the following section this trend towards rationalization is discussed in relation to investment activity.

The development of output and employment in manufacturing industry for the USA and Japan respectively clearly show different patterns of behaviour. There is no obvious change in trend in the

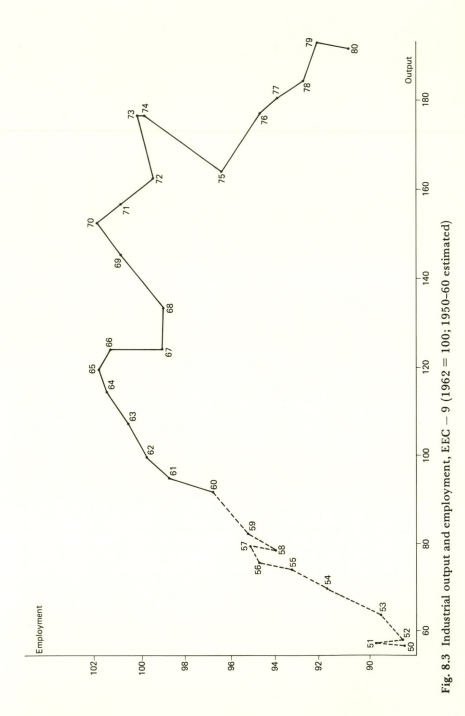

Fig. 8.3 Industrial output and employment, EEC − 9 (1962 = 100; 1950–60 estimated)

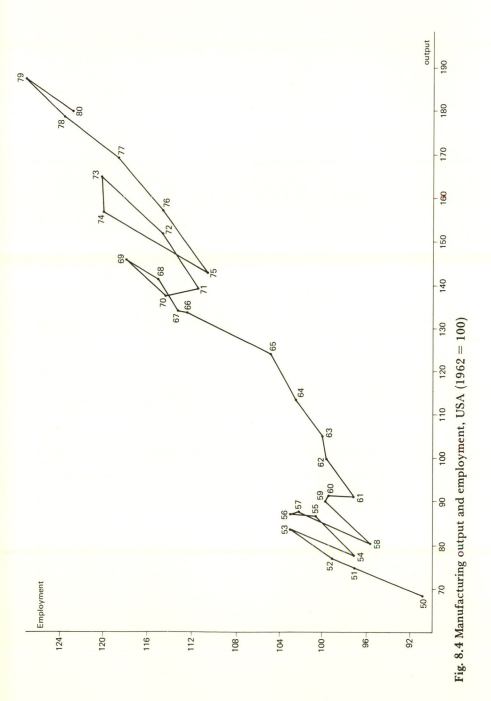

Fig. 8.4 Manufacturing output and employment, USA (1962 = 100)

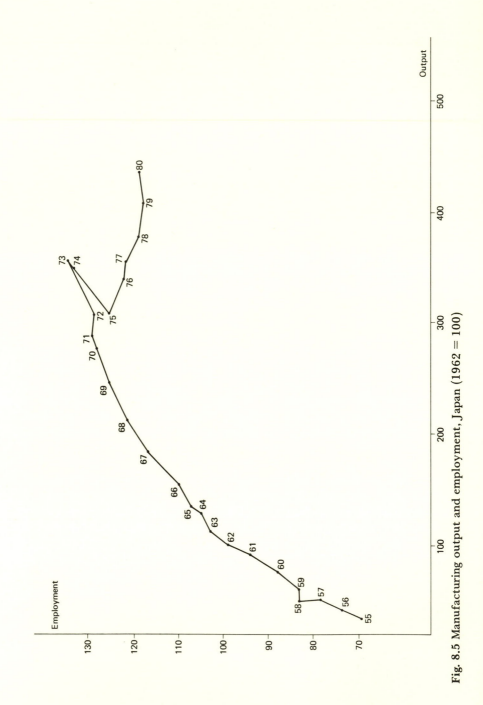

Fig. 8.5 Manufacturing output and employment, Japan (1962 = 100)

case of the USA, an observation confirmed by more detailed econometric analysis. For Japan, it appears that the 'rationalization'
period suggested by the EEC data for the latter 1960s has been
postponed to the latter 1970s, with employment declining somewhat since 1975, although output growth has been exceptionally
high throughout — note the important difference in the horizontal
scale of Figure 8.5 compared with Figures 8.3 and 8.4. For the
USA, a major shift in macro-economic trend over the last decade
is that in productivity growth, while for Japan the 'postponed'
onset of a reduction in manufacturing employment is of major
interest here. We shall return to discuss technological developments
in both these countries and their relationship to these trends in
Chapter 9.

8.3 Investment, capital intensity and employment

It has been suggested in Chapter 1 and elsewhere that the level
and nature of investment in fixed capital is of great importance
in relation to employment and unemployment, both in its role
as a component of aggregate demand and in extending the productive capacity of the economy. The purpose of this section is
to discuss some trends in investment activity over the last two
or three decades and their implications. It is not considered that
precise quantification is possible in this area, but it does appear
that important changes related to the 'long wave' have occurred
which carry some implications for 'investment needs' in the future.
   Table 8.6 presents average annual rates of growth of gross fixed
capital formation in manufacturing industry over five-year periods
for several countries. The figures in brackets show the annual growth
of employment in manufacturing in the corresponding periods.
The decades of the 1950s and 1960s are generally recognized as
a period of high investment, as implied by the rapid growth of non-
residential capital stock already shown in Table 8.3; Table 8.6
implies, however, that the growth of employment associated with
given rates of investment growth has varied substantially during the
1960s; of particular interest is the fact that in four of the six countries an *acceleration* in the growth rate of investment in the period
1965–70 was accompanied by a *reduction* in the growth rate of
employment compared with the previous five-year period. This
trend is also exemplified by Figure 8.6, which shows incremental
changes in employment in UK manufacturing plotted against investment (five-year moving averages in both cases). In the years following
the mid-1960s the annual change in labour input is negative and

Table 8.6 Investment and employment growth rates, manufacturing industry 1960–75 (employment in brackets)

	1960–65	1965–70	1970–75
France	6.9	9.6	3.5
	(1.1)	(0.5)	(0.4)
West Germany	3.2	5.2	−7.3
	(1.4)	(0.4)	(−1.9)
Italy	−3.3	11.3	−0.4
	(0.4)	(1.5)	(0.9)
Japan	6.0	21.5	−4.7
	(3.9)	(3.6)	(−0.5)
UK	3.7	4.2	−4.6
	(0.3)	(−0.5)	(−2.1)
USA	8.0	2.4	−2.2
	(1.1)	(1.7)	(−0.7)

Sources: Investment: OECD (1978) Statistical Annex 2; Employment: Brown and Sheriff (1979)

towards the end of the decade declines at an accelerating rate, despite continued high investment. This is in clear contrast to the 1957–62 period, where rather slower rates of investment growth are accompanied by overall positive and generally increasing increments in employment. The implication is that the onset of employment decline was not initiated by a reduction in the rate of growth of investment; however, the latter has, of course, declined during the 1970s, with further sustained reductions in the employed labour force.

Why did the high investment of the second half of the 1960s — amongst some of the highest growth rates in history for many countries — not generate large numbers of new jobs? One possibility is that changes in the pattern of final demand were such that the output of relatively 'capital intensive' sectors increased much more rapidly than that of labour-intensive sectors, so that quite independently of changes in technology or market structure new fixed capital was directed primarily towards sectors where little new employment is traditionally associated with a given increase in the capital stock. There seems to be no clear indication that such an effect was significant, however.

A further reason why high investment may not be accompanied by rapid employment growth is that there may be large-scale

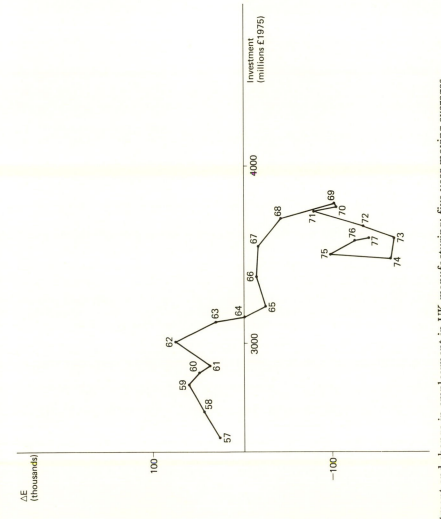

Fig. 8.6 Investment and change in employment in UK manufacturing: five-year moving averages

retirements of older capital equipment, which may be relatively labour-intensive. An increased rate of scrapping increases the investment rate 'required' to maintain or increase employment to an extent dependent both on the magnitude of the increased scrapping rate and on the difference between the capital intensities associated with new and obsolete equipment.

The rapid growth rates in capital intensity indicated by Table 8.4 suggest strongly that there is a considerable difference between the capital–labour ratios associated with recently installed equipment and that of older equipment on the margin of obsolescence. It follows that scrapping of the older equipment will in general necessitate an investment expenditure greater than the replacement value of the scrapped capital if employment levels are to be maintained.

Unfortunately, official statistics are of little value for testing the hypothesis of an increased scrapping rate. The 'perpetual inventory method', variations of which are used to derive the published capital stock estimates in all OECD countries, is based on the assumption that particular categories of capital equipment have a given fixed average lifetime; on this assumption, an increased scrapping rate would be suggested several years after a period of relatively heavy investment, when the fixed capital concerned has reached 'retirement age'. It is not inconceivable that a 'bunching' of retirements has been occurring for this reason. However, the *ad hoc* nature of the lifetime estimates used (and in particular the assumption that lifetimes are independent of economic and technological developments) mean that reference to capital stock statistics for evidence of changes in scrapping rates is of very limited value.

There are some indirect indications, however, that the rate of scrapping may have increased. For Germany, surveys carried out by IFO indicate a trend towards an increasing proportion of investment expenditures being devoted to 'replacement' or 'rationalization' — both of which imply that the installation of new fixed capital was accompanied by removal of old — rather than to an extension of capacity. As cited by McCracken *et al.* (1977):

IFO surveys of firms showed that in 1960 53 per cent of manufacturing companies cited expansion of capacity as the principal reason for their investment. By 1970, the proportion, as indicated by various IFO surveys of investment intensions, was 40 per cent and by 1977 it was 15 per cent.

Other data cited by McCracken *et al.* show that in several Western industrialized nations the share of productive investment devoted to non-residential construction declined between the early to mid-1960s and the mid-1970s, one interpretation of which is that more expenditure went on replacing machinery rather than in the development of new factory space.

It would perhaps be expected that, in times of low total (gross) investment, the 'replacement' or 'rationalization' component would be relatively high. Producers may always be interested in cutting costs and increasing productivity, and the act of replacing and updating the capital stock is a relatively low-risk decision, being independent of uncertain expectations of market growth. Thus this activity may plausibly occur at a roughly constant rate and hence form a higher (or lower) proportion of total investment as the latter falls (or rises). This pattern would be expected if the type of investment activity were determined by general economic conditions rather than being biased by a significant 'technology push' whereby producers may be induced to spend a 'disproportionate' fraction of their investable resources on rationalization in the light of a greater or lesser availability of profitable investment opportunities in that direction.

In so far as investment expenditures required per job created have increased, the most favoured general explanation (e.g., McCracken *et al.* 1977) would be in terms of a rise in input prices relative to output prices. Rapid wage increases, other things being equal, will tend to raise the relative price of labour-intensive products and the pattern of demand would be expected to shift towards capital-intensive sectors; it will also tend to accelerate scrapping as 'marginal' capital becomes unprofitable to operate more quickly; and it will tend to induce the adoption of capital-intensive techniques within individual sectors. However, our argument is that the forces of increased competition, greater concentration, economies of scale and other factors associated with the 'maturing' of the rapidly growing industries of the 1950s and early 1960s contributed significantly to the onset of 'rationalization' which, as suggested, began before the onset of the 1974–5 recession. Energy-saving measures following oil price increases, an increase in resources devoted to environmental protection, and (in some countries) rapidly rising real wages have doubtless exacerbated the trend, but their influence has probably been most severe in the mid to late 1970s.

Detailed discussions of the industries which grew around two of the major fourth-Kondratiev technological 'carriers' — electronics and synthetic materials — have been given earlier. It appears that both these industries were subject to increasing rationalization during the 1960s and in some respects their evolution provides a microcosm of the manufacturing industry-wide trends discussed here. However, there is no implication that all sectors of the economy have behaved in a similar way, or that where common trends do appear between industries, the reason is always the same. For example, rapid growth of capital intensity can occur in old as well as new (or 'maturing') industries. In an analysis of seventeen industrial sectors in France over the period 1950–70, Sautter (1979) asserts that 'growth of capital intensiveness tends to be fastest at the two extremes of the range of sector growth rates. The slow-growth sectors (coalmining, agriculture, textiles-clothing-leather, and iron, steel and ironmining) are, except for coalmining, sectors with falling employment and higher than average capital/labour substitution. The very rapid growth sectors (oil, water-gas-electricity, chemicals) are sectors where capital intensiveness is also increasing sharply but where employment is rising rapidly too'. Sautter points out that rapid increases in capital intensity combined with declining employment is often observed in industries adopting a defensive position with respect to foreign competition, as discussed in Chapter 7. In so far as decline would have been faster had these industries not accumulated capital more rapidly, their increased capital intensity can be interpreted as having had a job-saving effect, despite the statistical relationship apparently suggesting the reverse.

8.4 Capacity utilization and 'investment requirements'

It has been suggested that various factors have, over the course of the 'fourth long wave', reduced the employment generation effects of a given magnitude of fixed investment. The potential employment problems associated with this may, in recent years, have been exacerbated by the relatively slow growth of total investment itself since 1973 (see Table 8.7). This raises the possibility of severe supply side constraints to any return to 'full employment'.

Following the severe 1974–5 recession, a group was set up to analyse the difficulties facing the OECD economies and to make policy recommendations. Their report (McCracken *et al.* 1977) points out that 'the shocks of the recent past have led to widespread pessimism about whether sufficient investment will be forthcoming in future to generate the increase in capacity needed to provide enough jobs to make possible the progressive return to full employment' (para. 225). This group themselves considered that 'the evidence available to us concerning its actual significance and extent is equivocal' (para. 233).

It is interesting here to compare how capacity utilization with employment would provide evidence for the extent to which existing

Table 8.7 Annual percentage growth of gross fixed capital formation (whole economy-constant prices)

	1961–7 average (1)	1967–73 average (2)	1973–9 average (3)	1978 (4)	1979 (5)	1980 (6)
France	7.7	6.4	0.7	1.2	3.2	2.6
West Germany	2.5	5.7	1.3	5.2	8.3	4.2
Italy	2.9	4.3	−0.6	−0.1	4.4	8.0
Japan	11.4	13.3	2.4	9.6	8.3	0.1
UK	5.4	2.7	−0.6	3.5	−1.4	−4.5
USA	5.7	3.7	0.7	6.8	1.7	−7.4

Sources: Cols. 1–5: calculated from *OECD National Accounts-Main Aggregates* (1981, p. 87); Col. 6: *OECD Quarterly National Accounts Bulletin* and *EEC European Economy Annual Economic Report 1980–1981*.

spare capacity would be sufficient to restore manufacturing employment to earlier levels. In Figure 8.7 levels of capacity utilization are compared with manufacturing employment trends in a number of countries. As different techniques and definitions are used in estimating capacity utilization,* comparisons between countries cannot be made — only the trends over time within countries are of interest. As can be seen, utilization rates for the most recent years are comparable to those of some years in the early 1970s in all cases shown, while only in the USA has manufacturing employment not declined. In all countries shown except the USA, an at least partial recovery in utilization rates has not been matched by a recovery in employment.

In the case of the UK, the Confederation of British Industry, commenting on the most recent capacity utilization data, remarked that 'Interestingly, given the continuing decline in output, albeit more moderate than previously, the steadying of the spread of below capacity working probably implies that capacity itself has been reduced'. In the longer term, Figure 8.7(d) shows that capacity

*The level of capacity utilization may be estimated by various methods, falling into two main categories: 'judgemental' methods, which involve an industrial survey, and 'non-judgemental' methods, which make use of capital-stock estimates to compare actual production with potential production. In the former case the form of the question asked varies significantly in practice so that comparisons between countries are not possible. It is also believed that judgemental measures are subject to optimism or pessimism about what constitutes 'normal' output, depending on the general business climate (McCracken *et al.* 1977). 'Non-judgemental' methods depend, among other things, on assumptions made in the assembly of capital-stock series (e.g., an assumption of a fixed capital lifetime). The data presented in Figure 8.7 is taken from various national sources (as compiled by the OECD) and their reliability thus depends on the particular method used in each case.

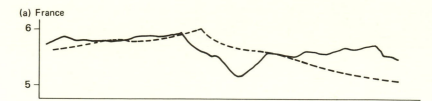

(a) France

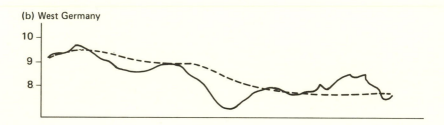

(b) West Germany

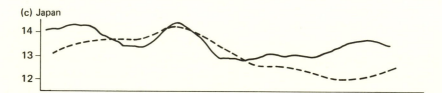

(c) Japan

(d) UK

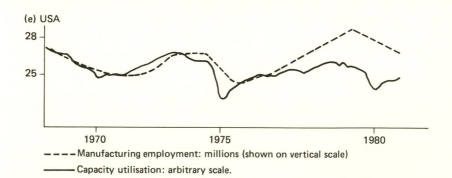

(e) USA

1970   1975   1980

– – – Manufacturing employment: millions (shown on vertical scale)
——— Capacity utilisation: arbitrary scale.

Fig. 8.7   Comparison of capacity utilization and employment trends in manufacturing, 1969–80, various countries. *Source*: OECD (1981).

utilization in 1977, for example, was roughly equal to that in the cyclic troughs of 1962 and 1967, while the number of employees was over 500 000 fewer in 1977 than in the earlier years. Similar pictures emerge for France, West Germany and Japan. While these data refer to manufacturing only, they suggest strongly that, given a revival of demand, employment growth could be constrained by a shortage of productive capacity. The conclusion is that developments since the McCracken report was compiled have strengthened the evidence for an 'investment problem' in many OECD countries. It is not intended here to consider the determinants of investment and the conditions that may be required for recovery; the prospects for the future in the context of the 'long wave' are discussed in Chapter 10.

In summary, this chapter has attempted to describe characteristics of post-war growth which exemplify some of the trends associated with phases of the 'long wave' described in Chapter 4. The 1950s and early 1960s represented recovery and boom, with rapid output growth accompanied in general by rapid increases in productivity and low unemployment. This was followed by the onset of 'stagflation', marked initially by continued rapid investment growth but with a slowdown in employment growth, rising prices and a decline in profit rates. The onset of output stagnation corresponded to substantial excess capacity, although the persistence of the current downturn may have led to considerable scrapping of older equipment in addition to a reduction in additions to capacity, leading to a 'shortage' of productive potential in manufacturing. The significance of this may be mitigated by a shift in demand from industry to services, although technological developments suggest considerable increases in the capital intensity of the service sector in the future, an issue to which we return later.

So far, most of the analysis has centred around more or less 'closed' country situations. Even when discussing the electronics and plastics case studies in Chapters 5 and 6, only little attention was paid to the international implications of new innovation, let alone 'clusters' of innovations. Yet, if there is one thing economic history, or history *tout court* brings out most dramatically, it is the close link between the possession of 'knowledge', with the focus on military technical superiority, but also social organizational adaptability, and the growth of countries, empires and even civilizations. Growth poles (as much as innovation poles) have switched around the world, from ancient Mesopotamia to Japan's present technological leadership.

Probably more than any others, the process of long-term economic growth which was set in motion in the UK in the eighteenth century with the Industrial Revolution was built around the possession of knowledge. It spread in 'waves' in the first half of the nineteenth century to the USA, France, Belgium, Switzerland and Germany and later diffused further to the European countries, Russia and a number of Latin American countries. The spreading process was sometimes extremely slow, hampered by structural and political factors (as in Argentina and Chile), or sometimes extremely rapid, based on organisational 'readiness' and active policy to promote the import of foreign technology and its internal diffusion (as in Japan).

The picture of the widely differing degrees of the international diffusion of new technology looks surprisingly similar to the picture of the wide variations in the degrees of adoption of new technologies by industries drawn earlier (see particularly Chapter 6). Similarly, when discussing the rise and fall of industries over the long wave in Chapter 7, one question which comes immediately to mind in this chapter is whether there is also some process of rise and fall of countries over the long wave and whether that process is in any sense related to the rise and fall of the main industry Kondratiev-carriers, in terms of export specialization patterns for example. Such a picture conforms with the argument of some long-wave writers (in particular van Duijn), that whereas individual countries data do not always indicate a clearly identifiable 'long wave' pattern, world data do generally speaking indicate more clearly such a pattern.

In this chapter we discuss some of the international aspects and implications of long waves and innovation 'clusters'. In a first section we briefly discuss the 'locus' of the major innovation clusters over the various Kondratievs. The change in locus from the UK to the USA, Germany and more recently Japan, pinpoints towards the crucial role of the international diffusion of technology, and knowledge in general, in the process of 'catching up' and what could be called 'Kondratiev-jumping'. In the second section we turn to some of the (unfortunately scarce) data which exist with respect to the origin of major innovations.

In the third section we discuss, in somewhat more detail, the process of 'catching-up' convergence growth, and put forward a number of 'long-wave/catching-up' arguments, which provide some explanation for the alternative process of divergence/convergence growth that seems to have characterized the economic development of most Western economies over the most recent Kondratievs.

9.1 The changing locus of economic growth and innovation

Economic history teaches us that there are substantial advantages in being the first country to implement new innovations. The reasons why this is so, are not always obvious. Traditional international trade theory, for example, points in the first instance to the gains from trade for the non-innovating trade partner in terms of cheaper imports. Even when innovation concerns new products, the early monopoly rents earned on the exports of the new products by the innovating country should disappear because of the international diffusion of technology (Krugman 1979). Yet, as the following brief historical description of the major innovation and growth loci over the various Kondratievs will make clear, successful '*appropriation*' by the innovating country (by which we mean the degree to which the technical knowledge can effectively be kept within the national boundaries of the innovating country), is the crucial factor for the innovating country to grow fast. This is even more important when one is talking about the sort of major innovations Schumpeter had in mind (steam power, railways and electrification), where the country that is able to quickly adopt and invest in the new technologies can be expected to grow faster than any other country that is slower or simply has no access to the new technologies.

In many ways, when it started in the UK the Industrial Revolution came at a time when there were practically no other countries that had the political and economic infrastructure for it to be assimilated. France was fighting its Napoleonic wars, the USA was only just becoming a political entity, and neither Germany, Italy, nor Belgium existed. It should therefore come as no surprise if the first Kondratiev was a purely British phenomenon. As Ray notices:

In the first half of the 19th century Britain dominated the, then still narrow, world economy and during the period 1820-1850 she produced two thirds of the world's coal, one half of iron, more than one half of steel, one half of cotton cloth, and forty per cent of all hardware. Britain has no competition . . . Britain, the innovator, was riding high on the Kondratiev wave of steam power, with more steam engines than the whole of the rest of the world put together. (Ray 1980, pp. 17-18).

Yet it is interesting to note how, throughout this period, the UK tried to limit the diffusion abroad of its technology (e.g., by prohibiting the export of machinery). It was primarily English technicians/entrepreneurs who, having emigrated to Belgium and France, established factories using the new technology in those countries, and later even created subsidiaries in Germany and Russia;* this led to the spread of the new technologies throughout Europe in the first half of the nineteenth century, and led to increased international competition for the UK.† The international spread of technology during the first Kondratiev downswing (the period 1814-42) was further amplified by the fact that the new technologies were relatively simple. As W. Arthur Lewis notes:

The new technology of the industrial revolution — for using steam, making textiles, mining coal and making iron — was ingenious, but simple. The first textile innovators made their machinery of wood, with metal for moving parts (Lewis 1978, p. 159).

In addition, most of the first Kondratiev innovations were in the first instance 'process innovations' (i.e., they consisted of making or transporting 'existing' goods, such as iron, textiles, clothes in new ways). It was consequently extremely difficult for the UK to 'appropriate' these technologies, rather they spread to:

any country which was already producing iron, textiles or clothes, or growing cereals — be it Sweden or Russia, Brazil, China, Japan or India. One should note,for example that India opened its first modern textile mills in the 1850s [1853], and its first modern ironworks in the 1870s (Lewis 1978, p. 30).

A similar picture can be drawn with regard to the second Kondratiev. Again, it started in Britain and was based on British innovations

---

*In 1815 Cockerill (located near Liège, in what was to become Belgium) set up probably the first ever subsidiary in Prussia (Franko 1976).
†An exception was Switzerland. Escher, a textile manufacturer introduced English machinery for his mill but then started to imitate it and initiated thereby the Swiss engineering industry.

(the first locomotive and the first railway). During the early phases of the second Kondratiev upswing, British industrial production rose more rapidly than anywhere else, yet by 1860 (when the upswing started to level off) the UK's growth rate started to lag behind that of a number of European countries such as Germany, Italy, Sweden, Switzerland and Belgium. In Ray's words:

The railway system had matured — the innovation having spent its initial driving force — earlier in pioneer Britain than in the countries which were 'followers'. However, at the end of this upswing Britain was still in the lead in many respects, unsurpassed in coal, iron (and the newcomer, steel) and cotton cloth production, but it was becoming clear that in other areas her leadership was being seriously challenged (Ray 1980, p. 18).

However, the second Kondratiev downswing sees the UK being overtaken as the most important growth and innovation world pole. This was linked both to the successful imitation and catching up of a number of European countries and the USA (primarily through import substitution and protectionism based on 'infant industry' ideas), and to the decline of some of the major industry carriers of the first Kondratiev (such as the cotton industry in which the UK had retained her lead). In the most important growth industries, primarily steel and metalworking, Britain — while having initiated some of the most important innovations (the Bessemer converter, Siemens' open hearth, in 1866, and the Thomas process, in 1878) — lost its innovative lead in exploiting these to Germany who overtook her in the production of iron, coal and steel by the end of the century.

The third Kondratiev sees the emergence of both the USA and Germany as major growth and innovation poles. The UK was clearly lagging behind with respect to the major innovations in relation to the third Kondratiev (electricity and automobiles). With respect to electricity it was primarily the USA (Edison's electric power steam central, 1882; Westinghouse's alternating current distribution of electric power, 1885) which took the lead; its production of electricity was about five times that of the UK by 1907, while its total industrial production was about the same (Ray 1980, p. 19). With respect to motor cars the lead was both European (Germany: Daimler, 1887; Krebs, 1892, and France: Panhard and Levassor, 1892-4) and American (Duryea, 1892). By 1910:

the largest car maker in Britain was Henry Ford, producing more cars than the next two largest firms combined. Thus, British entrepreneurs lagged in this major industry too, and this is likely to have had a considerable impact on the general development of a number of branches of the engineering industry in view of the requirements of automobile production in terms of machine tools and new machines — as well as some other newer and older industrial sectors such as rubber and instruments, or even textiles and timber (Ray 1980, p. 20).

The UK's situation over the third Kondratiev was in many ways exceptional, to the extent that it actually did not seem to take part in the third Kondratiev. The rate of growth of industrial production declined continually from 1873 onwards until World War I (i.e., over the greatest part of the third Kondratiev boom). There exists a large literature as to the reasons why, after 1883, the UK lost both her economic as well as her innovative lead so quickly and dramatically. Probably the most plausible explanation relates to the major causes underlying the present-day British decline (lagging behind in the provision of technical education and mass higher education, and the lack of linkages between scientific institutions and industry); these are all factors which were crucial to the rapid growth of both the USA and Germany, over the third Kondratiev upswing.

The gloomy picture that emerges from the UK's economic performance after 1883 is by all standards rather familiar: a low ratio of domestic investment with, as a consequence, a declining rate of growth of productivity, slow growth of exports with rapid growth of imports of manufacturers, and slow adoption of new technologies relative to competitor countries. This whole pattern is further exacerbated by the UK's free trade and 'self-chosen role as guardian of the gold standard' political position, compared with the import substitution–infant industry industrialization policies followed elsewhere. Yet it is the fact of the UK lagging behind, first in the adoption of her own innovations (e.g., in the steel industry referred to earlier), and later in the development of new innovation itself, that most obviously needs to be explained. We would agree with the core of Lewis's argument that, with the third Kondratiev upswing science had become increasingly important in contributing to new innovations; this was different from the earlier Kondratievs in which the role of science was extremely limited. Very often, the scientific principles of the new innovations had not yet actually been analysed (see, e.g., the steam engine and the lagging behind of thermodynamics); the UK seemed to be both particularly badly prepared and organized to cope with this fundamental change. According to Lewis:

It was not necessary for British entrepreneurs to be scientists; what mattered was that they should be receptive to science, and understand how to use scientists. There is no doubt that British ideology was at this time hostile to science, and even more so to industrial science. This was the great age of the public school (an offshoot of the railway, without which parents could not have moved so many children so many times a year) which, being based on the proposition that the English were the modern descendents of Periclean Athens, found little virtue outside the classics and religion. At this time too, the public service grew enormously, attracting the best brains into Parliament, the home civil service, the Indian civil service and the diplomatic service. Of course there were plenty of brains in the classes from which industrialists had typically been recruited. What gentlemen's sons thought or did is not relevant to us since industrialists had never been gentlemen's sons. What does matter is that the non-gentlemen's sons were going (as they always had) through a school system which held applied science and technology in contempt. This had not mattered in 1830, but it was crucial to industry in 1900, when industry was in need of an increasingly scientific base.

Consequently organic chemicals became a German industry; the motor car was pioneered in France and mass-produced in the United States; Britain lagged in the use of electricity, depended on foreign firms established there, and took only a small share of the export market. The telephone, the typewriter, the cash register and the diesel engine were all exploited by others. When after 1900 the economy was profitable, capital was exported instead of being invested at home. (Lewis 1978, p. 130).

By contrast, the USA was well organized and in many ways initiated the closer link between science and industrial technology (e.g., Edison's first research laboratory in Menlo Park, New Jersey, in 1879). Germany also succeeded in covering the gap between academic science and industrial progress, and this was to become crucial in the development of the major chemical innovations in the 1920s and 1930s (as described in Chapter 5). Yet, at the time of the third Kondratiev, the link between science and innovation was still relatively weak. Major innovations did not yet depend on new scientific discoveries or inventions in a crucial way; they consisted primarily of seeking ways to exploit technological inventions commercially. The progress of 'science' had indeed been dramatic over the whole nineteenth century, and had caught up with technological progress. Consequently there were plenty of technological opportunities that could be derived in a relatively straightforward manner from existing scientific knowledge. The major technological problem became how to develop new products for which the great unknown was the identification of potential demand.

The third Kondratiev major technologies are indeed different from the first and second Kondratiev ones in that there was much greater stress on *product* innovations (e.g., automobiles, telephones, gramophones, typewriters, cameras, airplanes, radios, etc.). These innovations were in many ways more easily 'appropriable'; they generated rapid output growth in the innovating countries, and created new demands all over the world for existing commodities:

copper for electric wiring, rubber for bicycle and motor car wheels, oil for the internal combustion engine, and nitrates for the wheatfields: it also created new trades in refrigerated meat and bananas. The population explosion coupled with rising incomes increased the demand for tea, coffee, cocoa, vegetable oils, raw silk and jute. (Lewis 1978, p. 30)

Consequently the innovations linked with the third Kondratiev upswing spread less (and more slowly) than any of the innovations of the two previous Kondratievs, partly because of their closer link with scientific knowledge and their more complicated nature, and partly because of their 'new product' nature which made it easier for the innovating countries to 'appropriate' the underlying technologies. While economic growth throughout the world was rapid during the third Kondratiev upswing, this was less because of the international diffusion of some of the major third Kondratiev technologies, but rather either because of increased international interdependence with rapid growth of raw materials exports and manufactured imports in the peripheral countries, including the so-called tropical countries, or because of the further diffusion of some of the major technologies of the first two Kondratievs in a number of late-industrializing countries. As a consequence, at the peak of the third Kondratiev there had been no newcomers to the very select group of 'technological leaders'.

With the coming of World War I, and World War II and the third Kondratiev downswing, the increased international world trade interdependence — brought about during the years 1883–1913 — was going to have disastrous effects for those peripheral countries who had become increasingly dependent on demand in the centre countries for their exports. First, as a result of World War I, and the massive capital (and human) destruction in Europe, the terms of trade moved against tropical products in the 1920s. Secondly, with the Great Depression, demand fell even more for peripheral commodities, and the terms of trade moved even further against tropical products. Investment came to a stand-still, and a large number of peripheral countries, primarily the independent Latin American countries, started a late but relatively successful industrialization import substitution development pattern. More than any other event, the third Kondratiev downswing, in particular the depression of the 1930s, brought home the crucial advantages of 'catching up', industrialization and the import of technology, and the dangers of relying excessively on exporting primary products to those peripheral countries who had neglected the domestic manufacturing opportunities over the third Kondratiev upswing.

The fourth Kondratiev saw the USA in the first instance but also

Germany (with respect to plastics) as the countries from which the major early Kondratiev innovations emerged. With World War II, the USA became the absolute technological leader and a huge technological gap came into existence after World War II. Many major new technologies characteristic of the post-war Kondratiev boom (such as televisions and computers) originated in the USA. Even with respect to scientific discoveries, the USA emerged as absolute leader, something it had not achieved during its previous Kondratiev technological leadership.

The massive capital destruction in Europe resulting from World War II, and the existing technological and scientific skills in most European countries created the conditions for rapid 'catching-up growth', where the mere existence of the huge technological gap between the USA and Europe was sufficient to trigger off a process of economic growth which was impressive by any historical standard. This process was accompanied by massive transfers of technology, under the form of licence agreements and the setting up of subsidiaries in Europe of American firms. While the forms of technology transfers were clearly designed to keep overall American control over the technology transferred abroad, the level of technological, educational and scientific skill in Europe led to its diffusion and successful imitation and innovation in many European firms. Thus, precisely when increasing concern was being expressed with regard to the technological gap between the USA and Europe (Servan-Schreiber 1967, OECD 1968), the gap was being closed at great speed.

The country that not only succeeded in regaining its position at the world technological frontier, but entered in great style the very select group of technological leaders, was of course Japan. The process of economic growth which took place in Japan during the fourth Kondratiev boom consisted in the first instance of a gigantic catching-up process during which most of the technological innovations of the fourth *and third* Kondratievs were being assimilated and diffused throughout the economy. It amounted to something which could be called Kondratiev-jumping or a 'super'-Kondratiev boom. Japan's success story has been the subject of much debate, which cannot be dealt with here. In our view, apart from the specific cultural Samurai background, what probably comes nearest to explaining Japan's success in assimilating and diffusing so rapidly foreign technology, was its specific concern with the transfer mechanism (i.e., its rejection of foreign direct investment as an efficient way of transferring foreign technology and its insistence on appropriating herself the imported technology). War destruction, US aid (including military aid which freed it from the need to re-build its military power), were no doubt crucial factors in enabling Japan to start a process of rapid catching-up, yet these factors do not provide any explanation for Japan's success in assimilating so rapidly foreign technology, and becoming herself a major technological power.

What brought Japan to such a specific 'technology-import' industrialization policy? According to Allen (1981) this seems to be more the result of historical accident rather than careful economic analysis:

A well-known Japanese economist, who takes a jaundiced view of the usefullness of his professional colleagues as advisers on policy, recently asserted that his country owes much of its postwar economic success to the fact that, at the end of the Second World War, most of the senior academic economists were Marxists. The postwar government, being conservative in temper, was naturally not disposed to turn to Marxists for advice. So it perforce fell back on bureaucrats and administrators whose economics had been learnt by experience. Some of these advisers were engineers who had been drawn by the war into the management of public affairs. A typical example is Saburo Okita, once of the Economic Planning Agency and recently Foreign Minister. They were the last people to allow themselves to be guided by the half-light of economic theory. Their instinct was to find a solution for Japan's postwar difficulties on the supply side, in enhanced technical efficiency and innovations in production. They thought in dynamic terms. Their policies were designed to furnish the drive and to raise the finance for an economy that might be created rather than simply to make the best use of the resources it then possessed.

The issue came to a head soon after the war when the government was considering how best to use its exiguous resources for industrial rehabilitation. It was natural that the more orthodox financiers, including the influential Governor of the Bank of Japan, should advocate policies that seemed consistent with Japan's factor-endowment of that time — a huge supply of under-employed labour, an extreme scarcity of capital, and out-of-date technology. The obvious candidates for development were the labour-intensive industries textiles, clothing, pottery, metal smallwares. It would be folly, so it was argued to ignore Japan's comparative advantages in such trades in the pursuit of goals only open to countries well supplied with capital and technical expertise. At first these views had an effect on what was done. The Bank of Japan saw to it that loans were denied (in 1951) for a project for building a new, up-to-date steel works. Sony was obliged to postpone its import of transistor technology because the officials in charge of foreign exchange licensing were doubtful both about the technology and about Sony's ability to make use of it. But, on the whole, the bureaucrats and their advisers at the Ministry of International Trade and Industry (MITI) prevailed. They repudiated the view that Japan should be content with a future as an underdeveloped country with low productivity and income per head. She should bend her energies, in their view, to building up an industrial system based on capital-intensive manufactures. It was true that she was at that time technically backward and short of capital. But there was no obstacle to her importing 'know-how' from the United States, and her large company of well-trained technicians would permit her to assimilate quickly what the West could provide (Allen 1981, pp. 68–69).

Whatever the exact reasons as to Japan's success, Japan has been the only major new-comer to the select group of technological leaders during the fourth Kondratiev, to the extent that both France and Germany took back a position which they had already acquired during the third Kondratiev upswing.

9.2 A brief statistical exploration

The historical picture sketched above conforms to what various data about the origin of major innovations, inventions or discoveries, teach us. In Figure 9.1 we present the ten-year moving average of the number of major *technical* inventions and innovations, selected from a list of 1012 major inventions, innovations and discoveries compiled by Streit (1949), by country of origin for the period 1755 to 1945.* Compiled shortly after World War II the list is unfortunately strongly biased towards the USA, and against Germany. In addition, its coverage of major innovations and discoveries for the period 1939–50 is extremely poor, because of the lack of retrospective vision in identifying major innovations at the time of compiling the list. Nevertheless, the picture which emerges conforms to what was said earlier. It should also be borne in mind that the data relate to the technical inventions or innovation *years* and have consequently to be lagged sufficiently (± thirty years) if one wants to take into account the full scale diffusion of the major technologies in terms of Kondratiev upswings.

The overall predominance of the UK with respect to the major innovations of both the first and second Kondratiev emerges most clearly. By 1850, however, the UK was overtaken first by the USA and then by France and Germany, where most of the major innovations that gave rise to the third Kondratiev originated. Unfortunately, with respect to the fourth Kondratiev, Figure 9.1 provides no information. By the end of the 1930s, however, the overall dominance of the USA (while exaggerated, certainly with respect to Germany) nevertheless conforms to what was said earlier with respect to the emergence of an 'abnormal' technological gap between the USA and Europe after World War II.

*See Table 9.1 for the number of major technical innovations and inventions, scientific discoveries and social innovations over the period 1750–1950 by Kondratiev phase. Figure 9.1 only relates to the technical innovations and inventions, and Figure 9.2 to scientific discoveries. Social innovations, while probably of crucial importance in terms of a country's organizational capacity to adapt have not been analysed, because Streit's list uses the concept of 'social' innovation in the broadest possible sense. Not only does the list include general and social achievements such as free principles of government, women's suffrage, and unemployment insurance, but also other 'human' achievements such as mass warfare, attributed to Napoleon, Mussolini's nationally applied fascism and Hitler's nationally applied national socialism.

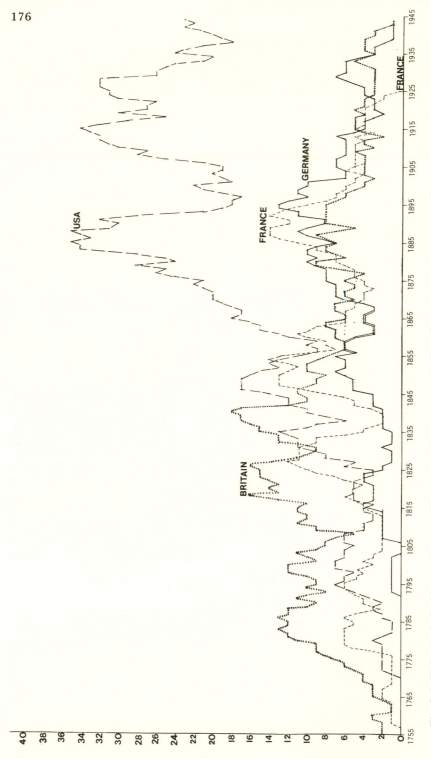

Fig. 9.1 Major technical innovations and inventions by country of origin (ten-year moving averages).

As to the other countries, their innovative activity in terms of major Kondratiev innovations remained relatively limited with the partial exception of Switzerland, Sweden and Belgium. They were less the initiators of some of the major 'epoch-shaping' innovations, than the successful 'fast' imitators who succeeded in reaping most of the benefits from the rapid progress of the world innovation frontier over the various Kondratiev upswings.

A totally different picture emerges from Figure 9.2, which indicates the ten-year moving averages of the number of major *scientific* discoveries by country of origin calculated on the basis of the figures in Table 9.1. The source is similar to Figure 9.1 and a similar bias might be expected. Yet, surprisingly enough, Figure 9.2 suggests a relatively weak scientific position for the USA over the whole of the nineteenth century, with its scientific lead only established in the mid-1930s; similarly, with reference to the UK, Figure 9.2 indicates a relatively weak position with respect to the first Kondratiev upswing (which seemed to have been dominated by France), Britain only emerging as the scientific leader by 1808, a position which she quickly saw challenged by Germany (1860s), who retained the scientific lead till the 1930s. The pattern emerging out of Figure 9.2 illustrates the changing role of science with respect to major innovations quite neatly: from an initial situation of very little influence (the first Kondratiev), to a situation of crucial importance: the 1930s. It also shows that, certainly in the nineteenth century, science (contrary to technology) was largely a 'non-appropriable' activity, which could relatively easily be transferred abroad — more specifically, 'scientific' leaders do not necessarily become 'technological' leaders.

With respect to the more recent period, Table 9.2 presents, for a number of years in the period 1883–1979, the number of patents granted in the USA to various countries and groups of countries in the world. As we have argued before (see Chapter 2), overall patent data do not allow one to separate 'major' innovations from minor improvement innovations. The purpose of Table 9.2 is only to give some indication of the overall importance of a number of non-USA countries (in particular Japan) over the most recent fourth Kondratiev. With the exception of Canada, whose proximity to the USA clearly overstates its importance, Table 9.2 confirms the picture of Figure 9.1 for the post-1880 period: the long-term decline of the UK, only temporarily halted by World Wars I and II; and the steadily growing importance of Germany as a technological power, which after World War II returned to its pre-war patenting level in less than nine years. It also shows the emergence of Italy,

Table 9.1 Major inventions, innovations, scientific discoveries and social innovations by country of origin (1750-1950)

Long-wave phase (dating from Kuznets 1940)	Technical inventions and innovations					Scientific discoveries					Social innovations				
	UK	USA	France	Germany	Other	UK	USA	France	Germany	Other	UK	USA	France	Germany	Other
1750-1786 pre-Kondratiev 1787-1800	20	5	8	0	1	6	1	6	1	9	5	4	7	1	1
1st Kondratiev: prosperity 1801-1813	14	7	6	1	2	3	0	12	3	1	5	3	3	0	1
1st Kondratiev: recession 1814-1827	10	6	2	2	2	15	0	9	1	0	1	1	0	0	0
1st Kondratiev: depression 1828-1842	21	5	8	4	2	1	1	5	4	8	1	1	1	0	0
2nd Kondratiev: recovery 1843-1857	22	14	11	2	2	4	0	4	8	1	1	7	0	1	2
2nd Kondratiev: prosperity	13	22	14	8	9	7	1	8	14	5	0	2	2	1	2

Period															
1858–1869	8	18	11	4	4	7	0	10	17	6	0	4	0	2	•
2nd Kondratiev: recession 1870–1884/5	11	37	9	15	7	2	4	6	8	5	2	3	3	0	1
2nd Kondratiev: depression 1886–1897	9	35	15	12	8	3	3	2	4	3	0	1	1	0	1
3rd Kondratiev: recovery 1898–1911	5	26	8	10	10	9	2	4	5	8	1	4	2	0	1
3rd Kondratiev: prosperity 1912–1924/5	6	47	4	5	3	3	3	4	5	6	1	3	0	0	4
3rd Kondratiev: recession 1925/6–1938	5	32	0	6	3	5	11	0	6	6	1	3	1	2	1
3rd Kondratiev: depression															

Source: Streit (1949). Classified into technical, scientific and social with the help of Joe Townsend.

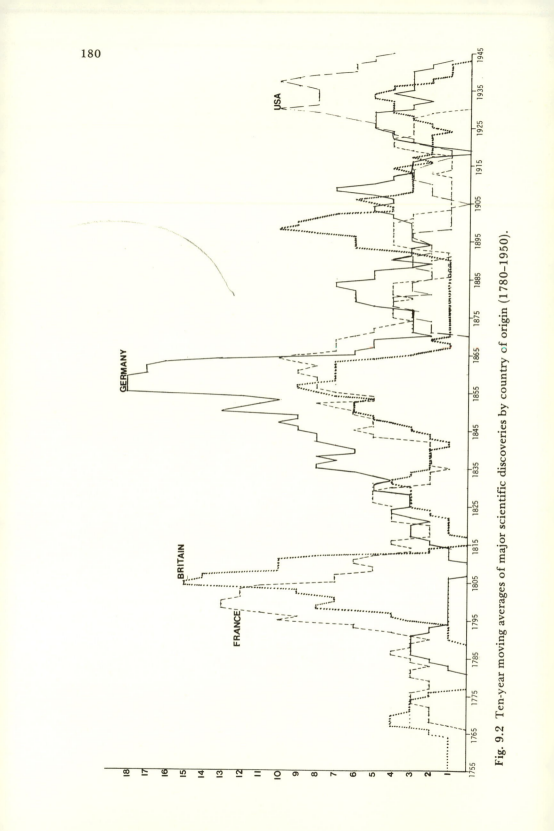

Fig. 9.2 Ten-year moving averages of major scientific discoveries by country of origin (1780–1950).

Table 9.2 Patents granted in the USA by country of origin (in percentages of all foreign patenting)

Countries	1883	1890	1900	1913	1929	1938	1950	1958	1965	1973	1979
Australia	1.11	1.20	2.33	1.97	1.96	1.18	1.54	0.60	0.94	0.92	1.12
Austria	2.62	3.37	3.36	3.99	2.47	2.91	0.48	1.12	1.16	1.02	1.19
Belgium	1.59	0.86	1.35	1.28	1.30	1.23	1.07	1.14	1.50	1.23	0.98
Canada	19.94	17.63	10.54	13.22	10.25	6.35	11.16	7.99	7.00	6.20	4.56
Denmark	0.56	0.38	0.46	0.67	0.71	0.71	1.36	0.74	0.74	0.70	0.56
France	14.22	8.46	9.79	8.07	9.76	9.23	15.54	10.36	10.90	9.38	8.46
Germany	18.67	21.47	30.72	34.02	32.36	38.18	0.57	25.60	26.40	24.25	23.87
Italy	0.24	0.29	0.92	1.31	1.91	1.43	0.86	3.02	3.38	3.39	3.14
Japan	0.16	0.10	0.03	0.45	1.40	1.51	0.03	1.93	7.43	22.10	27.69
Netherlands	0.24	0.29	0.75	0.47	1.57	3.38	8.10	5.71	4.15	3.03	2.80
Norway	0.32	0.14	0.49	0.74	0.71	0.54	0.95	0.61	0.42	0.42	0.43
Sweden	0.95	1.52	1.32	2.07	3.19	3.13	6.67	4.64	4.50	3.40	3.02
Switzerland	1.75	2.66	2.27	3.11	4.46	3.72	9.73	8.80	6.97	5.79	5.40
UK	34.55	36.15	30.52	23.29	22.23	22.70	36.00	23.45	20.62	12.56	10.07
East European Countries and USSR	0.40	0.67	1.49	1.19	1.62	1.61	1.23	0.55	0.89	2.53	2.76
NICs	0.40	1.19	1.12	1.21	1.03	0.90	1.41	1.31	1.71	1.36	1.45
Other	3.28	3.62	2.54	2.94	3.07	1.29	3.28	2.43	1.29	1.72	2.50

Source: Pavitt and Soete (1982).

the Netherlands, Sweden and Switzerland with larger shares in the twentieth century, and a somewhat increased share of France since World War II.

However, the most striking change in patent share relates to Japan. Its level of innovative activity, as measured by the number of US patents granted, remained among the lowest of all OECD countries until the 1920s; and it was only in the late 1950s and 1960s that, after having returned by 1957 to its pre-war patenting level, its share of foreign US patents started to grow very rapidly. In 1979, it was the major foreign country patenting in the USA, accounting for more than 27 per cent of total US patents of foreign origin.

As mentioned before, apart from Japan, there have been no new-comers over the fourth Kondratiev to the very select group of world innovators. The share of Eastern Europe and the USSR, while grossly under-estimated, has increased slightly since World War II. What are now called the newly industrializing countries have increased their share slightly, but it remains very small.

9.3 Some thoughts about catching up and long-wave economic development

As illustrated above, the introduction of the international dimension in our 'long wave' discussion highlights the crucial importance of the phenomenon of catching-up and the technology gap growth based on the international diffusion of some of the major Kondratiev technologies. That catching-up process is basically unrelated to the overall long wave up- and downswings, yet it is obvious that the possibilities for individual countries to set off on such a rapid growth process are, in the first instance, a function of their distance from the overall long-wave productivity or technological frontier. Thus, when the Kondratiev enters its downswing, apart from the technological leaders, countries that are near to approaching the technological frontier − particularly those countries that have greatly benefited over the upswing from a process of rapid 'catching-up' growth − will feel the effects of the slowdown in economic growth most directly. The productivity growth 'slowdown' in the 1970s of some of the major post-war 'catching-up' economies in the OECD area (Japan, Germany, France and most small European countries), fits this picture well. The productivity slowdown of the technological leader itself, the USA is more closely related to the overall Kondratiev downswing in this interpretation.

By contrast, other countries (e.g., the so-called newly industrializing countries, which are still in their early or 'booming' catching-up phase) will start to overtake the 'technologically mature' countries in terms of both output and productivity growth. From their view-point, the Kondratiev downswing may be seen as a welcome pause in the rate of advance of the technological or productivity frontier, opening up possibilities of becoming full members of the select group of economic and technological leaders in the future.

Figures 9.3 and 9.4 illustrate some of these catching-up phe-nomena. Figure 9.3 illustrates how the USA developed its tech-nology and productivity lead from around 1890, and how − after a process of 'divergence' from 1913 to 1950 − most of the major OECD countries started a process of rapid 'catching-up' growth, and came to or near the US productivity frontier in the later 1970s. The question of why the initial process of divergence came about, and what prevented the other countries catching up long before World War II, are undoubtedly linked with the overall long-wave pattern of the third Kondratiev. According to Maddison:

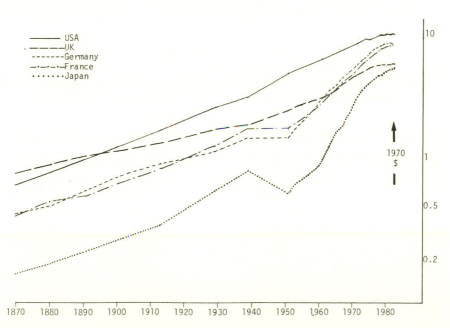

Fig. 9.3 Level and growth of GDP per man hour (1870–1980): USA, UK, Germany, France and Japan (in 1970 US dollars).
Source: Maddison (1979) and our updating.

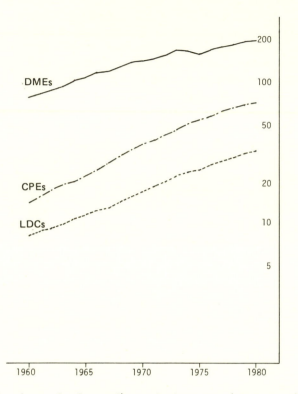

Fig. 9.4 Level and growth of manufactured value added (1960–1980): developed
market economies (DMEs), centrally planned economies (CPEs) and
less developed countries (LDCs) (1960 World total = 100). *Source*:
Calculated from UNIDO (1981).

The USA developed its productivity lead initially in the period from the 1890s
to 1913 at a time when its prospects were particularly bright because of its
great natural resource advantages, huge internal market and rapid population
growth. This fostered higher rates of investment than in Europe and a faster
growth of capital per employee. By 1913 the US productivity advantage over
the UK — the old leader — was about a quarter (Maddison 1979, p. 17).

Rapid growth in the USA in the third Kondratiev upswing was, of
course, also a function of its technological lead in the narrow sense
(i.e., the major third Kondratiev innovations originated in the USA).
Vis-à-vis the old technological leader (the UK), this led quite naturally to a process of divergence; *vis-à-vis* the other European major
countries, the picture is less clear. In these cases, it was (in the first
instance) the period of the third Kondratiev *downswing* with both
World Wars I and II that accentuated the process of productivity

growth divergence between the USA and the rest of the world. As a result 'In 1950, there was an unnatural degree of dispersion between the USA and most of the other countries' (Maddison 1979, p. 17) which 'triggered' off the process of rapid catching up growth.

Figure 9.4 illustrates the growth performance of both developed and developing countries over the period 1960–80. At the end of the 1960s, both the fourth Kondratiev downswing and the ending of the rapid catching-up growth phase of a number of developed countries, led to a marked reduction in the growth rate of industrial production, and manufactures in particular, of the developed countries. By contrast, the industrialization/catching-up process set in motion in a number of less developed countries, primarily the newly industrializing countries, was relatively little affected by the overall Kondratiev downswing, but continued to grow at about the same rate — if not even faster — over the period 1969–78. Whether this amounted to a 'catching-up' growth process is at present difficult to judge. Only a limited number of developing countries are involved in this process; the gap which they have to bridge is by any historical standard gigantic and new technologies have reached extreme levels of technical sophistication. The various factors that prevented them from 'catching-up' in the various previous Kondratievs, in particular the boom phase of the fourth Kondratiev — rapid population growth (thanks to the free and rapid diffusion of medical science and technology), integration in the world international and gradually liberated trade system (1913–39), difficulty in appropriating new technologies (the growth of multinational corporations) — are still all there. Nevertheless, it seems feasible and not too optimistic to imagine a world in the not too distant future which has a few more countries entering the group of technological mature societies.

The process of 'catching-up' growth has, of course, been described by many authors. The earliest versions (Rostow 1960, Gomulka 1971) focused primarily on the near 'mechanical' nature of economic growth once 'take off' had taken place; in many ways this was similar to the 'socio-mechanics' of the early Kondratiev theories. Once a country had achieved a certain level of development, growth was going to take place under any circumstances. This brought about a long chain of reactions (primarily the Latin American *dependencia* school, starting with Prebish, most UN writings on industrialization and technology transfer and various others). This line of discussion culminated in A. G. Frank's notions of the 'development of underdevelopment' and the 'accumulation of backwardness'.

This critique was useful in that it pinpointed the structural difficulties most developing countries had to cope with in achieving economic growth, and that it definitely broke with the notion of a freely available 'world-technology' which could be taken from the world shelves and transferred, and was sufficient to set into motion the process of industrialization. At the same time, however, and precisely at a time when rapid industrialization was taking place in a number of developing countries, it failed (in particular the more radical *dependencia* versions *à la* Frank) to explain the growth process itself.*

The 'catching-up' growth models accurately describe the process of convergence that has characterized the post-war growth pattern of many OECD countries (Cornwall 1977). They fail, however, to describe the 'divergence' process which lay at the basis of the initial 'technology gap'.† More recently Pavitt (1979, 1980a) has extended Cornwall's 'catching-up' model to include divergence growth based on 'product and systems innovations'. In Pavitt's model the departure from a convergence growth pattern is indicated by the fact that:

Entrepreneurial activity is no longer focused on responding to domestic demand by investing in borrowed and adapted foreign technology, nor is it focused on competing on the basis of price. It concentrates instead on seeking new product opportunities in all the industrially advanced markets, in market segments where quality may take precedence over price and on exporting equipment, technology, and skills to the newly industrialising countries. Its main competitive advantage is not geographical proximity to a richer expanding market but its ability to respond to multinational threats and opportunities emerging from changing tastes, technology, relative prices and competition (1979, p. 286).

This will eventually lead to a process of economic divergence. While interesting, Pavitt's model provides no explanation for the abrupt change in the process from catching up to divergence.

*For a more detailed critique see Soete (1981b).

†Their overall predictive power in terms of a convergence of all growth levels to population growth levels is particularly poor. For example, Gomulka (on the basis of his catching-up model) made some unfortunate predictions in relation to UK productivity growth for the mid-1970s and 1980s:

The UK industry is at present under the increasing influence of the technologically and organisationally more innovative, and apparently already more advanced Western Europe and Japan. So, in principle, one may expect a further though temporary rise in the UK's productivity growth from about 2.3 per cent in the '50s 4 per cent in the years 61–74 to 5 per cent or so, characteristic now for Western Europe and Japan, followed by a decline to the US rate of about 3 per cent per annum (Gomulka 1979, p. 191).

In reality productivity grew over the period 1975–1980/1 at an annual rate of less than 1 per cent.

This is, of course, where long-wave considerations become of crucial importance. While countries converge and draw nearer to the technological leader during the slowing down and the actual downswing of the Kondratiev, it is the recovery based on a 'cluster' of inter-related innovations originating in the old, or some new technological leader which will set the process of divergence in motion. The country at the technological frontier that is best prepared technologically, economically and above all organizationally, will probably become the new technological leader and a technological gap might start developing. Now, over the downswing of the fourth Kondratiev — with more and more countries drawing nearer the technological frontier — there is a period of intense technological competition between the major 'contenders' for technological leadership. This is accompanied by a continuing increase in autonomous research spending (as opposed to the preponderance of 'auxiliary-to-technology-import' research spending in the early catching-up phase) of the 'catching-up' countries, and more and more duplicative research. At the same time, the international diffusion of technology increases more rapidly, the various technological leaders competing against each other in domestic as well as foreign markets. The result is a decline in the rate of return to inventive activity and innovation. Even the smaller countries, seeing their productivity growth more and more restrained by the rate of advance of the technological frontier, will increase further their 'autonomous' research effort, in order to achieve technological leadership in those sectors in which they have either developed some technology or export comparative advantage. To the extent that the resulting increase in international technological competition reduces the rate of return to technology, most countries find themselves now in the paradoxical situation of spending more on 'autonomous' research while the price for the import of technology has actually slowed down. As each country's 'technological assets' represent less and less monopoly power, so that the monopoly rent it can earn on its technology exports starts declining, it will finally be the degree of technological success, the relative labour costs, the availability of skills, and the extent to which government support is provided that will determine which country(ies) will eventually emerge as the new technological leader(s) in the next Kondratiev upswing. This will be accompanied by a process of 'divergence-growth', which will last until the newly created technological gap is sufficiently large to provide a sufficient incentive for the 'lost-out' countries to catch up.

The problem of which country is best prepared to emerge as

the new technological leader in the next Kondratiev upswing is, of course, a highly speculative question. Yet all the evidence seems to point towards either the old technological leader, the USA, or the major newcomer Japan. Why we believe that the latter might eventually emerge as the new technological leader is as much related to Japan's technological capabilities and general human skill 'endowments' as to its capacity for rapid social organizational adaptations, and its success in 'appropriating' technology worldwide.

This book has been descriptive and analytical rather than prescriptive, and it is not easy to point to a simple set of specific policy conclusions. Indeed, fatalistic attitudes characterized some of the early Kondratiev literature, and paradoxically, could be made to fit some present-day economic thinking, apparently justifying the view that governments can do very little about the present recession, except to follow a policy of monetary constraint and to hope that investment does indeed prove to be sufficiently interest-elastic. They might also cling to the hope that depression might stimulate innovation. The analysis presented in this book does not, however, provide any support for such a crude technological determinism; rather the reverse. During such periods as the present recession there is a more urgent need to push forward the technological frontier and public policies have an essential role: but these are unfortunately the times when the (dynamic) allocative efficiency of markets is at its lowest; and when both industry structure, and the general state of expectations make 'entrepreneurs' risk-averse. All these call for an active public policy, which has a dimension largely lacking either in present day monetary constraint or post-Keynesian demand-stimulating policies. Before discussing the core of such a policy, let us briefly summarize the main conclusions from our analysis.

Throughout this book we have maintained that the emergence of new technologies and their assimilation in the economies of industrial countries, which is an extremely important aspect of long-term economic development, is not a smooth continuous process. Specifically, we have argued that technical change is extremely uneven over *time*; as between *industries* and broad sectors of the economy; and *geographically* as between regions and countries. The diffusion of clusters of technical innovations of wide adaptability is capable of imparting a substantial upthrust to the growth of the economic system, creating many new opportunities for investment and employment and generating widespread secondary demands for goods and services. Over *time*, however, these new 'technological systems' mature and their investment and employment consequences tend to change. The combination of standardization, growing capital intensity and scale economies means that the employment generated per unit of investment tends to diminish whilst profitability is eroded during the diffusion process. The 'compensation effects', which might mitigate these tendencies, operate only imperfectly and often with long delays. It has been argued (e.g., in Chapter 8) that much of the stock of capital equipment already in existence in manufacturing may be rather inflexible in terms of its employment potential (i.e., it is 'clay' rather than 'putty') and new vintages of equipment may require heavy investment.

If this is correct and the trend of technical change is strongly capital-using, in the sense that capital/output ratios tend to rise significantly, and labour per unit of capital decline strongly, the demand management problem could become more complicated. Demand would have to grow rapidly to keep up with productivity growth in manufacturing and a shift from a stable to a rising capital/output ratio requires that investment be rising as a fraction of GDP if demand growth is not to lead to 'capital shortage'. If the investment rate is too low, demand growth may be prevented from keeping employment growing, because employers do not have an adequate modern capital stock to provide enough jobs. Inflationary pressures (reflecting capital shortages) could then set in, even when there may be significant unemployment. Apparent surplus capacity may partly consist of vintages of equipment that are obsolete in terms of effective international competition. A 'capital shortage' problem could still hinder demand management, even if technical change was not particularly capital-using if increased investment was a low share of increased total demand. 'Capital shortage' unemployment is thus a real possibility in the short and medium-term in any substantial recovery process, unless a large increase in investment takes place.

Geographically and *between sectors*, the problems of unemployment in the older industrialized countries, especially in Europe and the North East of the USA, may be more serious than is commonly supposed. There are some long-term structural aspects associated with the decline of older industries and the maturity of several of the post-war new technological systems, so that in many industries (even in those where it proves possible to expand demand and increase investment) this may not necessarily lead to any big increase in employment, if these industries are to remain internationally competitive.

The extent to which public and private sector service employment, requiring relatively little associated investments, could compensate for these declining opportunities in manufacturing is a central one. If such employment generation is not associated with productivity increases and technical change in the tertiary sector as well, then it risks becoming a source of persistent inflationary pressure.

Millions of new jobs were generated in the USA during the 1970s and the vast majority of these were in the service sector. In the early 1970s many were in the public service both at Federal and State levels; but in the late 1970s (Rothschild 1981) most were in the private personal service industries, mainly in low wage and short working week occupations. It is tempting to infer from this experience that a combination of demand expansion with increased flexibility in wages and hours could resolve the problems and permit a rapid return to full employment in Europe as well as in the USA. As in the 1930s, downward flexibility of wages and the emasculation of union bargaining power by legislative or economic pressures are sometimes advocated as the decisive steps towards recovery.

However, before accepting such deceptively simple remedies it is essential to recognize that, in contrast to previous periods of recovery from recession, the 1975-9 apparent recovery — although more vigorous in the USA than in Europe — was not accompanied there by any significant productivity growth but *was* accompanied by a high rate of inflation by USA standards and by a loss of international competitiveness. These inflationary pressures led to the abandonment of expansionary policies in 1980 and to a renewed rise in unemployment. This suggests that simply to pursue a policy of demand expansion, even with low wage 'union-free' sectors as in the USA, could be a dangerous path for European countries with their far greater vulnerability to international trade and exchange rate pressures, and often higher rates of inflation.

But more crucially, and as argued before, in terms of our long-term analysis simple demand expansionary or stimulating policies, while useful in terms of avoiding some of the negative aspects of recessions such as large-scale unemployment, will actually be unable to provide the necessary impulse to the economic system in terms of 'tilting' it out of a long-term structural recession on to a higher path of economic growth. More than in any other period of long-term economic development, structural recessional/depressional periods call for a specific technology policy.

10.1 The role of technology policy

We have suggested that the promotion of major new technological systems and of productivity growth based on technical change may be an important means to help restore the economic health of the mature industrialized countries. This conclusion emerges from the earlier experiences of recovery from depression as well as from the more recent Japanese experience. Particularly important would be innovations of wide adaptability, which could lead to a combination of high productivity gains and greatly improved profitability.

Three sets of technology policies seem particularly relevant:

(i) Policies which aim directly at encouraging firms to take up radical inventions/innovations. They seem particularly relevant in those recessional/depressional phases, when private investment seems reluctant to go for these radical, but risky, innovations. They could be described as Schumpeter mark I policies, ranging from direct financial support, to various forms of indirect risk-taking support in order to promote the emergence of radical new innovations. The evidence that we have considered in this book indeed suggests that such radical innovations are often not immediately and obviously profitable. Only after a fairly long gestation period does 'take-off' occur. This means that during the gestation period positive and patient public policies of support, encouragement, experiment and adaptation can be extremely important. The computer is a particularly notable example. The unaided market mechanism is not enough.

Economists and a wider public are often rightly suspicious of governments putting public money into exotic new technologies because of disappointing past experiences with supersonic transport aircraft, and some military and nuclear projects. However, these were launched in quite different circumstances and often without much consideration of the economic and social aspects. Good technology policy requires considerable sophistication and co-operation between engineers and economists, as well as some good luck and well-informed public debate. The discussion initiated by Eads and Nelson (1971) and followed up by Pavitt and Walker (1976) and Nelson (1980a and b) is extremely important in this context. They have designated the circumstances in which public involvement can be useful and effective and emphasized in particular the extremely important distinction between 'exploratory development', which is relatively inexpensive and would often merit support from government sources, and full-scale commercial development, which is usually far more expensive and more seldom justifies the commitment of public funds to R and D. The expensive failures that several countries have experienced have largely stemmed from disregard of this basic distinction, from the power and prestige of specialized lobbies to influence government policies and from the associated absence of adequate public debate. Investment projects incorporating new equipment, and procurement of new products that meet advanced technical specifications and satisfy social requirements, may be a much more satisfactory form of public involvement at this stage than R and D subsidies.

There are many other ways in which public policies can help to promote the emergence of radical new technologies. Some of them are described in general terms in the OECD Report on *Technical Change and Economic Policy* (1980) and by Rothwell and Zegveld (1981). Support for fundamental research, for 'enabling' or 'fundamental' technology and policies designed to improve the coupling between the various parts of the science–technology–industrial system are all of critical importance.

(ii) However, the early stages of radical innovations do not have big economic effects. Only large-scale diffusion can have such effects and therefore a second set of policies aimed at improving the diffusion of existing, but still relatively new and radical, innovations throughout the various sectors is essential. We refer here in particular to 'sheltered' or highly concentrated sectors of the economy — 'closed' sectors that are not subject to the pressures of international competition, such as public utilities, public transport and large parts of the service sectors; and the typical state or private monopolies or 'cartelized' oligopolies, which may be unable or unwilling for a variety of reasons to promote the adoption of radical new technologies. Policies are needed to promote dissemination of information and diffusion of innovation throughout the economy. They range from direct support for small innovative firms* to general support or information programmes for the introduction of specific technologies (such as the relatively successful UK Micro-processor Applications Project MAP) which consisted of a fund available to firms in any industry to finance education, consultancy and development costs of products or processes using micro-electronic technology.

In fact, probably the most effective public policies with the greatest employment effects would be those involving programmes of public investment and the accompanying procurement of new products or use of new technology. For example, it might cost between $5 billion and $20 billion to 'wire up' most urban areas in a country the size of UK or the German Federal Republic in such a way that domestic consumers could take full advantage of the potential applications of information technology, including two-way communication with distribution networks ('tele-shopping')

*For overviews of various policy proposals with respect to small firms see Sweeney (1980), Rothwell and Zegveld (1981 and 1982).

financial services ('tele-banking') and so forth. Whilst this would not necessarily involve large-scale commitment of public funds, active government policies are essential in such areas as standards, definition of responsibilities and links with existing telecommunications networks. Moreover, such ambitious long-term policies should pay special attention to the needs of the education system, the health services and other social services in which direct public procurement and investment are essential, and which might otherwise be the Cinderella areas of social and technical change.

We also would include here policies aimed at eliminating some of the other bottlenecks with regard to new technologies, particularly in relation to so-called 'human capital'.

The skill shortages that inhibit the rapid diffusion of new technology systems are an obvious and important area of public involvement. Experience in Germany and Sweden in particular has demonstrated that imaginative and ambitious public programmes can complement the efforts of private firms in training and re-training the labour force, and minimize the adjustment problems of structural change especially for young people. Experience in Japan suggests that massive public investment in higher education combined with intensive training and re-training in industry may be even more effective. A much higher proportion of young Japanese attend higher education than in any European country now. There is scope for a variety of strategies tailored to national circumstances but there can be little doubt that a policy of 'intangible investment' in the expansion of education and training can be a useful aid to adaptation and a valuable 'counter-cyclical' investment strategy.

(iii) A third set of policies aims at improving the import and the internal diffusion of foreign technology. It is a policy that in the first instance has to convince local businessmen and managers, as well as government officials, that foreign technology in certain areas and at certain times might be more advanced or simply better than domestic technology; this seems to be particularly difficult to achieve in the case of the old technological leaders, such as the UK and the USA. Yet, as Japanese post-war experience shows, a deliberate policy towards the import of foreign technology, coupled with autonomous efforts to improve it, can be highly successful. Particularly for industries that are at some distance from the world technological frontier, such a policy seems extremely relevant, but even for technological leaders, active support in seeking and using the best available world technology is common sense. No country can be a technological leader in all areas, and all can learn from international experience. Perhaps the most important group in MITI is that which monitors world-wide technological developments and advises on possible future trends and their implications for Japanese industry.

10.2 The limitations of technology policy and of demand management

It would be a mistake to believe that technology and training policies *alone*, however well-conceived and executed, could extricate market economies from their present difficulties. Technology policy does not operate in a vacuum but in a very specific economic and political environment. The co-existence of high inflation with high unemployment is a major new phenomenon of the present Kondratiev downswing. Chesnais (1982) has reminded us that each successive wave of new technologies is diffusing in a very different social climate; Schumpeter discussed the new complications arising from the growing role of the state and of large oligopolies already in the recovery from the 1930s depression. He returned to these problems particularly in *Capitalism, Socialism and Democracy* (1943).

In Chapter 7 we have pointed to the further growth of industrial concentration since that time and to rigidities in price behaviour associated with that process. Schumpeter mark I is still important but Schumpeter mark II predominates. The inflexibility of wages has been a commonplace of the economic policy debate for more than half a century. There is much force in Kristensen's (1981) argument that almost all markets including commodities and capital as well as labour have become much less flexible than in the nineteenth century or the early part of this century. His analysis of inflation and unemployment is one of the most original and challenging in the recent debate and derives added force from his experience as a Finance Minister and as Secretary General of the OECD in the 1960s. After discussing the loss of price flexibility in many types of 'organized' market, he concludes:

. . . expansionary demand management may be able only temporarily to influence employment in a positive direction, and . . . the chances of lowering the rate of inflation through restrictive policies are poor. The possibilities of getting out of the present unhappy combination of inflation and unemployment must therefore be considered small or non-existent (p. 43).

He goes on to ask whether in these circumstances there is (as some monetarists suggest) a 'natural' and unavoidable rate of unemployment or of inflation and answers his own question:

In my view the word 'natural' is misleading in both cases. It is not in the nature of things that there must always be high rates of unemployment or of inflation. Rather with the present organisation of markets and of the public sector a strong tendency for both inflation and unemployment to be persistent features of modern economies seems to be built into the system. This does not prevent the performance of the system from being better in both respects if these systems of organisation are improved (p. 43).

In the concluding chapter of his book (in which he discusses the policy implications of his analysis) he raises the other fundamental question which is at the heart of the current debate:

If it is true that the economies have lost much of the self-adjusting capacity of former times, can we not go back to the old systems of the 19th century, when prices, wages and employment moved up and down but never got locked in an extreme situation for any length of time?

Again he gives an unequivocal answer:

There is no possibility of moving backward in history. Concentration of industry has been unavoidable because large enterprises are superior in branches where certain overhead costs, such as research and development, are important. The larger the market, the lower these costs per unit produced. For the same reasons multi-national corporations and international banking operations were bound to expand . . . Similarly it could not be avoided that wage earners united more and more. . . Finally the public sector has been bound to expand in modern societies. . . If we cannot go backward we must go forward. We must accept that markets are organised and increasingly so. And we must accept that the political authorities of national states have important roles to play, now supplemented by a number of international organisations (p. 134).

He concludes that the performance of the OECD economies can only be improved by the institutionalization of prices and incomes policies on a semi-permanent basis. Only such radical new departures could ensure a return to price stability and full employment.

Obviously, not everyone has been convinced by Kristensen's analysis and his powerful restatement of the radical Keynesian alternative to monetarism. But in so far as it relates to the role of innovation in the concentration process, to the growth of inflexibilities in markets, and to the role of government in policies for science and technology, we find his arguments persuasive. Whether economies which adopted his policy prescriptions could any longer be reasonably described as 'market' economies is an open question, but in any case we have already made clear that our approach to the behaviour of the economic system, like that of Schumpeter and Kristensen, is essentially a historical one which tries to take account of institutional and social change, as well as the specific features of successive waves of new technology. The widespread exploitation of information and communication technologies in the tertiary sector, as well as the introduction of new energy technologies, would in our view inevitably involve a substantial role for public authorities, as well as large and small private firms.

10.3 Technical change, employment and inflation

Whether or not one accepts these Schumpeterian/neo-Keynesian prescriptions, there remain two very fundamental issues which will confront the OECD economies in the next twenty years. The first is the extent to which technical innovation and its diffusion may alleviate the inflationary pressures which are present everywhere even during periods of recession. The second is the response of the work force to technical change, to 're-deployment', 're-training' and possibly temporary or prolonged unemployment. The problems are in our view closely related, since only a high rate of technical change can generate the type of productivity increases that can increase real — as opposed to money — incomes and a high rate of technical change increasingly requires the active participation of the work force at all levels. We shall return to this last point at the end of the chapter, but we first consider the question of the rate of technical change and its bearing on the rate of increase in real wages.

Clearly a zero-growth economy, or still more a declining economy, cannot satisfy the aspirations of its citizens for a rising standard of living or can only satisfy the ambitions of a small fraction of them at the expense of the rest with all the potential for social conflict which that implies. A rapidly growing economy, on the other hand, risks generating inflationary pressures which may become unacceptable if a 'free' labour market operates in conditions of strong demand for labour and acute skill shortages.

Whether or not the alternative solution of some kind of incomes policy can be made to 'stick' depends, in part, on the social learning process that we have gone through in the past quarter of a century. However, it would at least be more likely to stick if part of the deal were a significant, though small, improvement in the real income of the majority of the population on a relatively regular basis. Such an improvement is possible only through a high rate of technical change. If the responsibility for that technical change and the rewards for its effective implementation are more widely diffused and the whole process is more widely understood and appreciated, then this can only benefit the implementation of growth-oriented but anti-inflationary policies. 'Self management', 'worker-participation' and co-operative ownership are all important in this context.

10.4 Lessons from Japan?

One of the factors hindering a high degree of public involvement and understanding is itself the prevalence of high levels of unemployment. It is hardly surprising that workers, who may be declared redundant when opportunities for new employment are rather scarce, should be lukewarm about labour-displacing innovations. Not surprisingly too, workers in Japanese corporations who enjoy a high degree of job security ('lifetime employment') and are thoroughly informed about projected changes in the techniques in the plant in which they work, are apparently very co-operative in implementing such changes. It would be surprising if the continuing high rate of technical change and productivity increases in Japan were not related to these factors.

Several other features of Japanese economic performance stand out from the (now numerous) attempts to explain her success. First, as mentioned earlier, the Japanese government (and particularly MITI) has had a very deliberate and long-term policy towards the promotion of technical innovation in the Japanese economy. This was never simply a policy of carbon-copy imitation, but has always involved a combination of seeking and using the best available technology in the world with autonomous Japanese efforts to improve it. (In fact, in the early 1980s, Japan has ceased to be a net 'importer' of technology and has become a net 'exporter', in other words her technological balance of payments has moved into surplus.) Long-term technical considerations were always given priority. For example, when traditional economic and financial theorists argued in the early 1950s that no special measures should be taken to promote the car industry, as Japan could not compete successfully in this industry, they were over-ruled. Secondly, Japanese policy has been eclectic in its choice of means to promote the desired high rate of technical innovation but has seldom hesitated to use the power of government and the associated financial weapons to promote the R and D, investment and structural change on the scale believed to be necessary. Thirdly, Japanese investment, in both physical equipment and education and training, has been extraordinarily high and a close connection has been established between the introduction of new products and processes and training programmes at all levels in industry. This appears to involve very close co-operation with unions at plant level. Finally, it has recently been Japanese policy to promote the introduction of entirely new technologies (in global terms) on a small-scale demonstration basis in order to acquire experience, diffuse the results in industry and sustain economic progress. It is for a combination of these reasons that we have suggested that Japan may lead the next expansionary upswing of the world economy.

The Japanese experience confirms that, contrary to superficial appearances, it is easier to maintain full employment with a high rate of technical change and *vice versa*. When productivity growth is rapid, wage increases can be more easily absorbed without increasing production costs and without diminishing international competitiveness. Even if the direction of technical change is adverse, levels of investment as high as that in Japan can still generate further increases in employment, albeit more slowly. Economies of scale and the associated productivity gains are more easily attained under a high growth and high investment regime.

Other countries devoted relatively greater resources to R and D in the post-war period but in none of them were the links between economic policy and technology policy so close as in Japan. In the USA, the USSR and the UK a very large part of the highest quality scientific and engineering resources were locked into the military/nuclear/space sectors, which played virtually no part until very recently in the Japanese system. The tendency in the US and UK especially was to treat technology policy as a special aspect of defence with heavy prestige overtones, rather than as an integral part of economic policy and industrial policy. Some other European countries have been rather more successful but none of them have been anything like as successful as the Japanese.

But is the 'Japanese' path a practicable or a desirable one for other OECD countries to follow? Some features of it are obviously very attractive — in particular the combination of high investment, high rates of technical change and relatively low levels of unemployment. Many people would also consider the rather high degree of job security at least in the large-company sector as another attractive feature. However, the difficulties must also be taken into account: Japan has certainly not escaped inflation, although its consequences have been sufficiently dampened to sustain international competitiveness. Japan's success has been relative to her international competitors, some of whom had to be losers if trade generally grows more slowly. The 'two-tier' nature of the economy and the social and environmental strains imposed must also be considered. Japan, too, has its low-wage service sector in which working conditions are still poor and security of employment less than in the modern large firm industrial sector or in government. Social services are still far from adequate and housing is poor. Clearly it would also be difficult for other OECD countries to imitate some features of

Japanese society, which depend on very strong cultural traditions and behaviour patterns. Japan is far from being a Utopia and certainly cannot offer complete solutions to the most fundamental problems confronting all industrial economies in the last part of this century. However, her experience does offer some useful pointers on how to achieve high rates of growth and high levels of employment and technical change even in adverse world conditions.

10.5 Concluding remarks

To combat inflationary pressures with deflationary policies and high unemployment puts many of the social and political achievements of the entire post-war period at risk.

We find the acceptance of a high level of unemployment as a form of restraint on wage pressures, whether temporary or permanent, socially unacceptable and we believe it to be politically impracticable over any extended period. There has to be a better way to treat human beings in the twentieth century. We are not suggesting that technology policies alone can be successful in solving the fundamental social, political and economic problems that go far beyond the scope of this book. However, we do believe that well-conceived technology policies are a vital ingredient of any strategy designed to combat the twin crises of unemployment and inflation.

APPENDIX A: GLOSSARY

Invention The first idea, sketch or contrivance of a new product, process or system, which may or may not be patented.

Innovation The first introduction of a new product, process or system into the ordinary commercial or social activity of a country.

Diffusion The spread of an innovation through a population of potential users, with or without modification, both nationally and internationally.

Basic innovation as defined by Mensch (1975) Innovations that create a new market and a new branch of industry.

Basic inventions The inventions that lead to basic innovations.

Radical inventions and innovations The most important inventions and innovations, which may typically require a new textbook to describe them, may give rise to a change of technique in one or more branches of industry, or may themselves give rise to one or more new branches of industry.

Major inventions and innovations The next most important category of inventions and innovations, which may typically give rise to new products and new processes in existing branches of industry, to many new patents and to new chapters in revised editions of existing texts on technology.

Minor and incremental inventions and innovations Small improvements to existing products and processes, which may be patented but often will not, and which may merit brief mention in the technical literature, but often will not.

Master patent as defined by Baker (1976) The first patent to be economically viable in relation to a specific invention.

Key patents as defined by Baker (1976) The most important patents in relation to a specific invention.

Method of computation of life cycle curves
(note relevant to Figure 5.4)

The relative weight of each specified subject [papers, patents, production, researchers etc.] seems to be a more important indicator than the absolute number . . . in itself. . . . The changes with time of the relative weights of each item thus obtained are plotted on a diagram. For the items which seem to have a life cycle character, parameters a, b, c of the following function are calculated using the least squares method:

$$ y = \frac{c}{1 + a(x - b)^2} $$

Where y is the relative weight of the specific item in the year x. . . . In this type of function the parameters can be interpreted in the following way: a is the coefficient related to the breadth (time duration) of a life cycle curve; b is the time at which relevant researches [etc.] are executed . . . ; c is the relative weight of the relevant publications [etc.] at the peak of its life cycle curve. . . . This function therefore makes it easier to express the characteristics of the life cycle in a visual way . . . (K. Yamada and E. Otaki 1971, p. 355).

REFERENCES

Abernathy, W. J. and Utterback, J. M. (1975) 'A dynamic model of process and product innovation', *Omega,* 3 (No. 6).

Abernathy, W. J. and Utterback, J. M. (1978) 'Patterns of industrial innovation', *Technology Review,* 80 (June–July 1978).

Abernathy, W. J. and Utterback, J. M. (1979) 'Dynamics of innovation in industry', in Hill, C. T. and Utterback, J. M. (eds), *Technological Innovation for a Dynamic Economy,* Oxford, Pergamon.

Allen, G. C. (1981) 'Industrial policy and innovation in Japan', in Carter, C., (ed.), *Industrial Policy and Innovation,* London, Heinemann, pp. 68–87.

Baker, R. (1976) *New and Improved — Inventors and Inventions that have Changed the Modern World,* London, British Museum Publications.

Barr, K. (1979) 'Long waves: a selected annotated bibliography', *Review,* 11 (No. 4, Spring), pp. 675–718.

Beveridge, W. M. (1st Baron Beveridge) (1944) *Full Employment in a Free Society,* London, Allen and Unwin.

Blattner, N. (1979) 'On some well-known theoretical propositions on the employment effects of technical change', published in Schweiz. Nationalfonds, Nationales Forschungsprogramm 'Regionalproblems', Informationsbulletin der Programmleitung, Nr. 2, Bern.

Braun, E. and MacDonald, S. (1978) *Revolution in Miniature — The History and Impact of Semiconductor Electronics,* Cambridge University Press.

Brown, C. J. E. and Sheriff, T. D. (1979) 'De-industrialisation in the UK: Background Statistics', National Institute of Economic and Social Research, Discussion Paper No. 23.

Burns, A. F. (1934) *Production Trends in the United States Since 1870,* NBER, New York (reprinted 1950).

Chesnais, F. (1982) 'Schumpeterian recovery and the Schumpeterian perspective — Some unsettled issues and alternative interpretation', in Giersch H., 1982, op. cit.

Chrystal, K. A. (1979) *Controversies in British Macroeconomics,* Oxford, Philip Allan.

Clark, J. (1980) 'A model of embodied technical change and employment', *Technological Forecasting and Social Change,* 16 (No. 1, January), pp. 47–65.

Clark, J., Freeman, C. and Soete, L. (1981a) 'Long waves and Technological developments in the 20th century', in Petzina, D. and Van Roon, G. (1981), op. cit.

Clark, J., Freeman, C. and Soete, L. (1981b) 'Long waves, inventions and innovations', *Futures,* 13 (No. 4), pp. 308–22 (special issue).

Coombs, R. W. (1981) 'Innovations, automation and the long-wave theory', *Futures,* 13 (No. 5), pp. 360–70.

Cooper, C. M. and Clark, J. A. (1982) *Employment, Economics and Technology: The Impact of Technical Change on the Labour Market,* Brighton, Wheatsheaf (forthcoming).

Cornwall, J. (1977) *Modern Capitalism: Its Growth and Transformation*, Oxford, Martin Robertson.

Davies, S. (1979) *The Diffusion of Process Innovations*, London, Cambridge University Press.

Delorme, J. (1962) *Anthologie des Brevets sur les Matières Plastiques*, Vols. I–III, Paris.

Diebold, J. (1952) *Automation: The Advent of the Automatic Factory*, New York, Van Nostrand.

Dosi, G., (1981) 'Institutions and markets in high technology industries: An assessment of government intervention in European microelectronics', in Carter, C. F. (ed.), *Industrial Policies and Innovation*, London, Heinemann.

Dosi, G. (1982) 'Technical paradigms and technological trajectories — A suggested interpretation of the determinants and directions of technical change', *Research Policy* (forthcoming).

Dow, J. C. R. (1964) *The Management of the British Economy 1945-1960*, NIESR and Cambridge University Press.

Downie, J. (1958) *The Competitive Process*, London, Duckworth.

Dubois, J. H. (1967) 'Plastics industry', *SPE Journal*, June.

Duijn, J. J. van (1979) *De Lange Golf in de Economie: Kan Innovatie ons uit het dal Helpen?* Assen, Van Gorcum.

Duijn, J. J. van (1981) 'Fluctuations in innovations over time', *Futures*, 13 (No. 4), pp. 264-75 (special issue).

Dupriez, L. H. (1947) *Des Mouvements Economiques Généraux*, Louvain.

Eads, G. and Nelson, R. R. (1971) 'Government support of advanced civilian technology', *Public Policy*, 19 (No. 3), pp. 405-27.

EEC (Commission of the European Communities) (1978) *EEC Maldague Report I — Sectoral Change in the European Economies from 1960 to the Recession*, Brussels.

EEC (Commission of the European Communities) (1980) 'European economy', *Annual Economic Review, 1980-81*, November 1980.

EEC (Commission of the European Communities) (1981) *Eurostatics: Data for Short-term Economic Analysis*, No. 10, Edition B, Eurostat, Luxembourg.

Fels, R. (ed.) (1964) Abridged edition of Schumpeter, J. A., *Business Cycles: a Theoretical, Historical and Statistical Analysis of the Capitalist Process*, New York, McGraw Hill.

Forrester, J. (1981) 'Innovation and economic change', *Futures*, 13 (No. 4), pp. 323-31 (special issue).

Franco, L. G. (1976) *The European Multinationals: a Renewed Challenge to American and British Big Business*, London, Harper and Row.

Freeman, C. *et al.* (1963) 'The plastics industry: A comparative study of research and innovation', *National Institute Economic Review*, No. 26, pp. 22-60.

Freeman, C. *et al.* (1968) 'Chemical process plant: Innovation and the world market', *National Institute Economic Review*, No. 45.

Freeman, C. (1971) *The Role of Small Firms in Innovation in the United Kingdom Since 1945: Report of the Bolton Committee of Inquiry into Small Firms*, Research Report No. 6, London, HMSO.

Freeman, C. (1974) *The Economics of Industrial Innovation*, Harmondsworth, Penguin.

Freeman, C. (1979) 'The Kondratiev long waves, technical change and unemployment', in *Structural Determinants of Employment and Unemployment* Vol. 2, Paris, OECD, pp. 181-96.

Freeman, C., (ed.) (1981) 'Technical innovation and long waves in world economic development', *Futures*, 13, (No. 4), special issue.

Freeman, C. (1982) 'Some economic implications of microelectronics', in Cohen, D. (ed.) *Agenda for Britain: Micro Policy Choices*, Oxford, Philip Allan.

Frowen, S. (ed.) (1982) *Controlling Industrial Economies*, Basingstoke, Macmillan (forthcoming).

Garside, W. R. (1980) *The Measurement of Unemployment: Methods and Sources in Great Britain 1850-1979*, Oxford, Basil Blackwell.

Garvy, G. (1943) 'Kondratiev's theory of long cycles'. *Review of Economic Statistics*, 25 (No. 4), pp. 203-20.

Gelderen, J. van (1913) 'Springvloed: Beschouwingen over industriële ontwikkeling en prijsbeweging', *De Nieuwe Tijd*, 18 (Nos. 4, 5 and 6), April–June 1913.

Giersch, H. (1979) 'Aspects of growth, structural change and employment — A Schumpeterian perspective', *Weltwirtschaftliches Archiv*, 115 (4), pp. 629-51.

Giersch, H. (1982) Ed. *Proceedings of Conference on Emerging Technology; consequences for Economic Growth, Structural Change and Employment in Advanced Open Economies*, Tübingen, J. C. B. Mohr (forthcoming).

Glismann, H. H. *et al.* (1980) *Lange Wellen Wirtschaftlichen Wachstums*, Kiel Discussion Paper No. 74.

Gold, B. (1980) 'On the adoption of T.I. in industry: Superficial models and complex decision processes', *Omega*, 8 (No. 5), pp. 505-16.

Gold, B. (1981) 'Technological diffusion in industry: Research needs and shortcomings', *Journal of Industrial Economics*, March, pp. 247-69.

Golding, A. M. (1972) 'The semi-conductor industry in Britain and the United States: A case study in innovation, growth and the diffusion of technology', DPhil. Thesis, University of Sussex.

Gomulka, S. (1971) *Inventive Activity, Diffusion and the Stages of Economic Growth*, Skrifter fra Aarhus Universtets Økonomiske Institut nr. 24, Aarhus Institut of Economics.

Gomulka, S. (1979) 'Britain's slow industrial growth — Increasing inefficiency versus low rate of technical change', in Beckerman, W. (ed.) *Slow Growth in Britain: Causes and Consequences*, Oxford University Press, Clarendon, pp. 166-93.

Gourvitch, A. (1940) *Survey of the Economic Theory on Technological Change and Employment*, New York, Augustus M. Kelley (reprinted 1966).

Graham, A. K. and Senge, P. M. (1980) 'A long wave hypothesis on innovation', System Dynamics Group, Sloan School of Management, MIT (mimeo).

Hamberg, D. (1966) *R and D: Essays in the Economics of Research and Development*, New York, Random House.

Heertje, A. (1977) *Economic and Technical Change*, London, Weidenfeld and Nicolson.

Hessen, B. (1931), 'The social and economic roots of Newton's *Principia*, in Bukharin, N. (ed.) *Science at the Crossroads: Papers from the Second International Congress of the History of Science and Technology, 1931* (revised edition), London, F. Cass.

Hicks, J. (1974) *The Crisis in Keynesian Economics*, Oxford University Press.

Hill, T. P. (1979) *Profits and rates of return*, Paris, OECD.

Hufbauer, G. (1966) *Synthetic Materials and the Theory of International Trade*, Cambridge, Mass., Harvard University Press.

ILO (International Labour Organisation) (1980) *Yearbook of Labour Statistics 1980*, Geneva, ILO.

Jahoda, M. (1982) *The Social Psychology of Employment and Unemployment*, Cambridge University Press (forthcoming).

Jewkes, J. *et al.* (1958) *The Sources of Invention*, London, Macmillan; revised edition, 1969.

Kaldor, N. (1966) *Causes of the Slow Rate of Economic Growth of the United Kingdom*, Cambridge University Press.

Kaletsky, A. (1981) 'Crowding out recovery', *Financial Times*, 14 August, p. 12.

Katz, B. G. and Phillips, A. (1982) 'Government, technological opportunities and the emergence of the computer industry', in Giersch, H. (1982), op. cit.

Keirstead, B. S. (1948) *The Theory of Economic Change*, Toronto, Macmillan.

Keynes, J. M. (1936) *The General Theory of Employment, Interest and Money*, New York, Harcourt Brace.

Klein, B. H. (1977) *Dynamic Economics*, Cambridge, Mass., Harvard University Press.

Kleinknecht, A. (1981) 'Observations on the Schumpeterian swarming of innovations', *Futures*, 13 (No. 4), pp. 293-307 (special issue).

Kondratiev, N. (1925) 'The major economic cycles', *Voprosy Konjunktury*, 1, pp. 28-79; English translation in *Review of Economic Statistics*, 18 (November 1935), pp. 105-15; reprinted in *Lloyds Bank Review, No. 129* (1978).

Kristensen, T. (1981) *Inflation and Unemployment in Modern Society*, New York, Praeger.

Krugman, P. (1979) 'A model of innovation, technology transfer, and the world distribution of income', *Journal of Political Economy*, 87, pp. 253-66.

Kuczynski, J. (1976) *Vier Revolutionen der Arbeitskräfte*, Berlin, Akademie der Wissenschaft.

Kuhn, T. (1962) *The Structure of Scientific Revolutions*, Chicago University Press.

Kuznets, S. (1930) *Secular Movements in Production and Prices*, Boston, Houghton Mifflin.

Kuznets, S. (1940) 'Schumpeter's business cycles', *American Economic Review*, 30 (No. 2, June), pp. 257-71.

Kuznets, S. (1954) *Economic Change*, London, Heinemann.

Levin, R. (1980) 'Towards an empirical model of Schumpeterian competition' (mimeo), Yale.

Lewis, W. A. (1978) *Growth and Fluctuations 1870-1913*, London, Allen and Unwin.

Lieberman, M. G. (1978) 'A literature citation study of science-technology coupling in electronics', *Proceedings of the IEEE*, 66 (No. 1, January), pp. 4-13.

Lukoff, H. (1979) *From bits to bits — A personal history of the electronic computer*, Portland, Robotics Press.

McCracken, P. *et al.* (1977) *Towards Full Employment and Price Stability*, Paris, OECD.

Maddison, A. (1979) 'Long run dynamics of productivity growth', *Banca Nazionale del Lavoro , No. 128*, pp. 3-43.

Maddison, A. (1980) 'Western economic performance in the 1970s: A perspective and assessment, *Banca Nazionale del Lavoro Quarterly Review, No. 134*, pp. 247-89.

Mahdavi, K. B. (1972) *Technological Innovation: An Efficiency Investigation*, Stockholm, Beckmans.

Mandel, E. (1972) *Der Spätkapitalismus*, Suhrkampf, Frankfurt. See also Mandel (1981).

Mandel, E. (1975) *Late Capitalism*, New Left Books. Revised English edition of Mandel (1972).

Mandel, E. (1980) *Long Waves of Capitalist Development: The Marxist Interpretation*, Cambridge University Press.

Mandel, E. (1981) 'Explaining long waves of capitalist development' *Futures*, 13 (No. 4), pp. 332-8 (special issue).

Mansfield, K. (1961) 'Technical change and the rate of imitation', *Econometrica*, 29 (No. 4), pp. 741-66.

Matthews, R. (1968) 'Why has Britain had full employment since the war?' *Economic Journal*, 78, pp. 555-69.

Mensch, G. (1971) 'Zur Dynamik des Technischen Fortschritts', *Zeitschrift für Betrielswirtschaft*, 41 (No. 5), pp. 295-314.

Mensch, G. (1975) *Das Technologische Patt: Innovationen uberwinden die Depression*, Frankfurt, Umschau; English edition (1979), *Stalemate in Technology: Innovations Overcome the Depression*, New York, Ballinger.

Mensch, G. (1981) 'Long Waves and Technological Developments in the 20th Century: Comment', in Petzina, D. and Van Roon, G. (1981), op. cit.

Metcalfe, J. S. (1981) 'Impulse and diffusion in the study of technical change', *Futures*, 13 (No. 5), pp. 347-59.

Mitchell, B. R. (1975) *European National Statistics 1750-1970*, London, Macmillan.

Mowery, D. and Rosenberg, N. (1979) 'The influence of market demand upon innovation: A critical review of some recent empirical studies', *Research Policy*, 8, pp. 102-53.

Neisser, H. P. (1942) 'Permanent technological unemployment', *American Economic Review*, 32 (No. 1), pp. 50-71.

Nelson, R. (1980a) 'Balancing market failure and government inadequacy: The case of policy towards industrial R and D', Working Paper No. 840, Yale.

Nelson, R. (1980b) 'Parsimony, responsiveness and innovativeness as virtues of private enterprise: An exegesis of tangled doctrine', Institute for Social and Policy Studies, Yale.

Nelson, R. and Winter, S. G. (1977) 'In search of useful theory of innovation', *Research Policy*, 6, pp. 36-76.

Nevins, A. and Hill, F. (1954) *Ford* (2 vols). New York, Scribner's.

Nishiyama, S. (1981) 'The impact of new electronic technologies', in Giersch, H. (1982), op. cit.

OECD (1968) *Gaps in Technology, Analytical Report*, Paris, OECD.

OECD (1978) 'Revised study on past and present trends of industrial investment', DSTI/IND/28.3 (Rev), Paris (mimeo). See also Statistical Annex.

OECD (1979) 'Sectoral shifts and productivity growth' in *Economic Outlook*, 25 (July).

OECD (1980) 'Productivity trends in the OECD area', CPE/WP2(79)8, Paris (mimeo).

OECD (1981) *Main Economic Indicators*, Paris, OECD.

Pareto, V. (1913) 'Alcuni relazioni fra la stato sociale e la variazoni della prosperita economica', *Rivista Italiana di Sociologia* (September–December), pp. 501–48.

Pasinetti, L. L. (1981) *Structural Change and Economic Growth: A Theoretical Essay on the Dynamics of the Wealth of Nations*, Cambridge University Press.

Pavitt, K. (1979) 'Technical innovation and industrial development – 1 The new causality', *Futures*, 11 (No. 6), pp. 458–70.

Pavitt, K. (1980a) 'Technical innovation and industrial development – 2 The dangers of divergence', *Futures*, 12 (No. 1), February, pp. 35–44.

Pavitt, K. (1980b) (ed.) *Technical Innovation and British Economic Performance*, Basingstoke, Macmillan.

Pavitt, K. and Soete, L. (1980) 'Innovative activities and export shares: Some comparisons between industries and countries', in K. Pavitt (ed.), (1980b), op. cit.

Pavitt, K. and Soete, L. (1982) 'International differences in economic growth and the international location of innovation', in Giersch, H. (1982), op. cit.

Pavitt, K. and Walker, W. (1976) 'Government policies towards industrial innovation: A review', *Research Policy*, 5 (No. 1), pp. 11–97.

Peck, M. J. and Wilson, R. (1982) 'Innovation, imitation and comparative advantage: The case of the consumer electronics industry', in Giersch, H. (1982), op. cit.

Petzina, D. and Van Roon, G. (eds) (1981) *Konjunktur, Krise, Gesellschaft: Wirtschaftliche Wechsellagen und soziale Entwicklung im 19. und 20. Jahrhundert*, Stuttgart, Klett-Cotta Verlag.

Phillips, A. (1971) *Technology and Market Structure*, Lexington, Lexington Books.

Phillips, A. and Katz, B. G. (1982) 'Government, economies of scale and comparative advantage: The case of the computer industry', in Giersch, H. (1982), op. cit.

Prais, S. J. (1976) *The Evolution of Giant Firms in Britain*, NIESR, Cambridge University Press.

Price, D. (1965) 'Is technology historically independent of science?', *Technology and Culture*, 6 (No. 4).

Ray, G. (1980) 'Innovation in the long cycle', *Lloyds Bank Review No. 135*, pp. 14–28.

Rees, A. (1957) 'The meaning and measurement of employment and full employment', in *The Measurement and Behaviour of Unemployment*, New York, National Bureau of Economic Research, Princeton University Press.

Rosenberg, N. (1976) *Perspectives on Technology*, Cambridge University Press.

Rostow, W. W. (1960) *The Stages of Economic Growth*, Cambridge University Press.

Rothschild, E. (1981) 'Reagan and the real America', *New York Review of Books*, 5 February, pp. 12–18.

Rothwell, R. and Zegveld, W. (1981) *Industrial Innovation and Public Policy: Preparing for the 1980s and the 1990s*, London, Frances Pinter.

Rothwell, R. and Zegveld, W. (1982) *Small and Medium Sized Manufacturing Firms: Their Role in the Economy, Innovation and Employment*, London, Frances Pinter (in press).

Sahal, D. (1980) 'The network and significance of technological cycles', *International Journal of Systems Science*, II, No. 8, pp. 985–1000.

Salter, W. (1960) *Productivity and Technical Change*, Cambridge University Press.

Samuelson, P. A. (1939) 'Interactions between the multiplier analysis and the principle of acceleration', *Review of Economic Statistics*, 21 (No. 2), pp. 75–8.

Samuelson, P. A. (1981) 'The world's economy at century's end', *Japan Economic Journal*, 10 March, p. 20.

Sautter, C. (1979) 'Investment and employment on the assumption of slower growth', in *Structural Determinants of Employment and Unemployment*, Vol. II, Paris, OECD, pp. 131–64.

Scherer, F. M. (1965) 'Firm size, market structure, opportunity and the output of patented inventions', *American Economic Review*, 55 (No. 5), pp. 1097–125.

Schmookler, J. (1966) *Invention and Economic Growth*, Cambridge, Mass., Harvard University Press.

Schumpeter, J. A. (1912) *Theorie der Wirtschaftlichen Entwicklung*, Leipzig, Duncker and Humboldt.

Schumpeter, J. A. (1934) *The Theory of Economic Development*, Cambridge, Mass., Harvard University Press (translation of Schumpeter, 1912).

Schumpeter, J. A. (1939) *Business Cycles: A Theoretical, Historical and Statistical Analysis of the Capitalist Process* (2 vols.), New York, McGraw-Hill.

Schumpeter, J. A. (1943) *Capitalism, Socialism and Democracy*, New York, Harper and Row (Second edition, 1947).

Sciberras, E. (1977) *Multinational Electronics Companies and National Economic Policies*, Greenwich, Conn., Jai Press.

Sciberras, E. (1980) 'Technical innovation and international competitiveness in the television industry' (mimeo), Science Policy Research Unit.

Scitovsky, T. (1978) 'Market power and inflation', *Economica*, 45, pp. 221–33.

Scott, M. F. G. with Laslett, R. A. (1978) *Can We Get Back to Full Employment?*, Basingstoke, Macmillan.

Servan-Schreiber, J. J. (1968) *The American Challenge (Le Défi Americain)*, London, Hamilton.

Soete, L. L. G. (1978) 'International competition, Innovation and Employment', paper prepared for Workshop on the Relationship between Technical Development and Employment, Paris, 13–14 November 1978; Six Countries Programme on Aspects of Government Policies Towards Technological Innovation in Industry.

Soete, L. L. G. (1979a) 'Firm size and inventive activity: The evidence reconsidered', *European Economic Review*, 12, pp. 319–40.

Soete, L. L. G. (1979b) 'The measurement of inventive activity and its relation to firm size', paper prepared for the Sixth Conference of the European Association for Research on Industrial Economics, Paris, 9–12 September 1979 (mimeo).

Soete, L. L. G. (1980) 'The impact of technological innovation on international trade performance: The evidence reconsidered', OECD Science and Technology Output Indicators Conference, Paris, September 1980.

Soete, L. L. G. (1981a) 'A general test of technological gap trade theory', *Weltwirtschafliches Archiv*, 117, (4), pp. 638–66.

Soete, L. L. G. (1981b) 'Technological dependency: A critical view' in Seers, D. (ed.) *Dependency Theory: A Critical Reassessment*, London, Frances Pinter, pp. 181–206.

Solow, R. M. (1969) *Growth Theory – An Exposition*, Oxford, Clarendon Press.

Sorrentino, C. (1981) 'Unemployment in international perspectives' in Showler, B. and Sinfield, A. (eds) *The Workless State – Studies in Unemployment*, Oxford, Martin Robertson.

Streit, C. (1949) *Union Now: A Proposal for an Atlantic Federal Union of the Free* (2nd edn.), New York, Harper.

Sturmey, S. G. (1958) *The Economic Development of Radio*, London, Duckworth.

Sweeney, G. P. (1980) 'New entrepreneurship and the smaller firm', paper prepared for Six Countries Programme on Aspects of Government Policies towards Technological Innovation in Industry, Limerick, 9–10 June (mimeo).

Terleckyj, N. (1963) *Research and Development: Its Growth and Composition*, Studies in Business Economics no. 82, New York, National Industrial Conference Board.

Terleckyj, N. (1974) *The Effects of R & D on Productivity Growth in Industry*, Washington, NPA.

Tinbergen, J. (1981) 'Kondratiev cycles and so-called long waves: The early research', *Futures*, 13 (No. 4), pp. 258–63 (special issue).

Townsend, J. *et al.* (1981) 'Science and technology indicators for the UK: Innovations in Britain since 1945' (mimeo), Science Policy Research Unit.

Trevithick, J. A. (1980) *Inflation – A Guide to the Crisis in Economics*, Harmondsworth, Penguin (2nd edn).

UNIDO (1981) *World Industry in 1980*, Biennial Industrial Development Survey, ID/269, New York, UNO.

United States Federal Trade Commission (1977) 'The semiconductor industry: A survey of structure, conduct and performance', Staff Report, Bureau of Economics.

US Department of Commerce, Bureau of the Census (1980), *Statistical Abstract of the United States 1980*, Washington.

Vandoorne, M. and Meeusen, V. (1978) 'A clay-clay vintage model as an approach to the problem of structural unemployment in Belgian manufacturing', Rijksuniversitair Centrum, Antwerpen, Working Paper 7808 (mimeo).

Walsh, V. *et al.* (1979) 'Trends in invention and innovation in the chemical industry', Report to SSRC (mimeo), Science Policy Research Unit.

Wiener, N. (1949) *The Human Use of Human Beings: A Cybernetics Approach*, New York, Houghton Mifflin.

Weinstock, U. (1964) *Das Problem der Kondratiev-Zyklen*, Berlin.

Wragg, R. and Robertson, T. (1978) *Post-war Trends in Employment, Productivity, Output, Labour Costs and Prices by Industry in the United Kingdom*, Research Paper No. 3, London, Department of Employment.

Yamada, K. (1982) 'A study on time lag between life cycle of a discipline and resources allocation', *Research Policy* (forthcoming).

Yamada, K. and Otaki, E. (1971) 'Life cycle of basic research – An approach to the quantitative analysis of R and D activity', Research Policy, 1, p. 355.

Zwan, A., van der (1979) 'On the assessment of the Kondratiev cycle and related issues', Centre for Research in Business Economics, Erasmus University, Rotterdam (mimeo).

INDEX

Automation, 119
Automobile industry, rapid growth of, 71-2

Babbage, Charles, 107
Bakelite, 87-8
BASF (drugs company), dyestuffs inventions, 93
Bayer (drugs company), dyestuffs inventions, 93
'Black economy', 4, 5
Business cycles, and innovation, 57; influenced by exogenous events, 19; Schumpeter's failure satisfactorily to explain, 22; see also Kondratiev long wave theory; Schumpeter, Joseph

Calculators, 102
Capital, growth in stock per hour worked, 150; trends in output-capital ratio, 150-1; utilization, 162-5.
Capital accumulation, cyclical trends, 27; effects on employment, 30; factors affecting, 29; fixed, inflexibility of, 190; slowdown, 150
Capital destruction, 134; effect of scrapping older equipment, 160; in World War II, and consequent growth, 128
Carothers, W. H., 88
Chilton's Law, 99
Computers, 102; commercial market late to develop, 108; entry of large electronics firms, 110; history of industry, 107-9; leading to automation, 119; main beneficiaries, 122, 124; surge of publications, 109; time scale for adoption, 122

Demand, deficiency related to unemployment, 9, 10, 11; evolution over time, 140; management, 190; related to income, 140; saturation effects on industries, 32
Depression, acting as accelerator on innovation, 51-7; cyclical nature, 18-19; declining industries fate during, 134; Mensch's theory of innovation 'bunching' during, 44-63; to generate revival of growth, 80; see also Recession

Diffusion theory, 68-70; models, 68, 69; see also Innovation: diffusion process
Drug industry, 82; effect of patents on profits, 78; invention upsurge precedes investment, 85; post-war innovative surge, 20
Du Pont Corporation, 69; lead in synthetic materials, 93; profits from nylon, 93
Dyestuffs industry, importance of German inventions, 93

Economic growth, catching up and divergence, 186-7; dangers of relying on exports, 172; degree of 'inevitability', 185-6; equilibrium, 31-5; fastest growing sectors, 128; international (1960-78), 185; post-World War II, 20, 148-9; variations between industrial sectors, 127
Economies of scale, 75
Economy, effects of innovation clusters, 67; structural change, 31-4, as continuous process, 131
Electronics industry, 101-26; definition difficult, 102-3; growth of firms, 143; life-cycle, 101-2; major semi-conductor product innovations, 115; post-war innovative surge, 20; rationalisation, 161; slow commercial acceptance, 125; 'swarming' period, 116, 118
Employment, affecting attitudes to innovation, 198, 199; and capital utilisation, 162-5; and investment levels, 157-60; and new technology, 75; creation less associated with fast-growing sectors, 135; future influence of technical change, 197; in declining sectors, 128; in electronics industry, 118, 120-1; in new technology sectors, 128; in plastics industry, 99; related to output, 153; structural problem in industrialised countries, 190; see also Full employment; Labour force; Unemployment
Engel's Law, 140
Entrepreneurs, adopting new inventions, 41

Factors of production, constraints related

85
88